BIFURCATION
AND
CHAOS IN SIMPLE
DYNAMICAL SYSTEMS

BIFURCATION AND CHAOS IN SIMPLE DYNAMICAL SYSTEMS

J. Awrejcewicz
Technical University
Institute of Applied Mechanics
Lódź, Poland

World Scientific
Singapore • New Jersey • London • Hong Kong

Published by

World Scientific Publishing Co. Pte. Ltd.,
P O Box 128, Farrer Road, Singapore 9128
USA office: 687 Hartwell Street, Teaneck, NJ 07666
UK office: 73 Lynton Mead, Totteridge, London N20 8DH

Library of Congress Cataloging-in-Publication Data

Awrejcewicz, J. (Jan)
Bifurcation and chaos in simple dynamical systems/by J. Awrejcewicz.
p. cm.
Includes bibliographical references.
ISBN 9810200382
1. Bifurcation theory. 2. Chaotic behavior in systems. 3. Boundary value problems—Numerical solutions. I. Title.
QA372.A97 1989
515'.35—dc20 89-22433
CIP

Printed in Singapore by JBW Printers and Binders Pte. Ltd.

PREFACE

This book consists of three chapters. In the first chapter, analytical methods to solve the Hopf bifurcation problem are presented. These methods are based on the perturbation and harmonic balance techniques well known in nonlinear dynamics. One parameter Hopf bifurcation, biparameter Hopf bifurcation and bifurcation into quasiperiodic orbits in autonomous systems are considered. Then the analytical approach for detecting Hopf bifurcation solutions in Duffing, Mathieu and Mathieu-Duffing oscillators are demonstrated.

The second chapter includes a numerical approach to systematically study the behaviour of mechanical and bio-mechanical systems governed by the non-autonomous and autonomous nonlinear differential equations. The procedure presented is based on solving a boundary value problem using the shooting method. The behaviour of the dynamical systems are traced along variations of the chosen parameters. Observation of the evolution of the characteristic multipliers gives information about stability and possibly further branching of the investigated solutions. In the case of nonautonomous systems, periodic orbits are traced and the transition to chaos when one of the multipliers crosses the unit circle at +1 or −1 (period doubling) are discussed and illustrated. An example is given for chaos, which has appeared after the simultaneous passage of a pair of multipliers through the unit circle of the complex plane. An autonomous system of three nonlinear differential equations which governs the oscillations of human vocal cords is considered in some detail. Steady state solutions, Hopf bifurcations and the branches of periodic orbits which emanate from bifurcation points are calculated. Tracing the evolution of characteristic multipliers allows the observation of alternations of stability and possibly branching to other periodic or quasiperiodic motion.

In Chapter 3, the attention concentrates on formulating the analytical condition for Hopf bifurcation of the periodic orbit or the Hopf type bifurcation of the quasiperiodic orbit which lead to nonlinear algebraic bifurcation equations. Three examples show the occurrence of chaotic orbits after bifurcations.

The contents of this book are considerably influenced by my experience at the seminar at the Polish Academy of Sciences in Warsaw (1984–1987) and at the seminar organized by the Stefan Banach International Mathematical Center (September 19–December 18, 1986). I express my thanks to Prof. W. Szemplinska-Stupnicka for his knowledgable comments on part of the results presented in Chapters 1 and 3.

A majority of results presented in this book are based on a research project supported by the Alexander von Humboldt Foundation. I would like to acknowledge the many useful discussions with Prof. E. Brommundt, Prof. D. Ottl and H. Staben regarding the material given in Chapter 2. I wish to express my sincere thanks for the hospitality of the Institute of Technical Mechanics of the Carolo Wilhelmina University of Braunschweig, where part of the work presented in this book was carried out.

Finally I acknowledge a very fruitful collaboration with R. Klopp from the University of Waterloo, who has helped avoid many difficulties with the style and presentation of my work.

JAN AWREJCEWICZ
Braunschweig, Spring 1989

CONTENTS

CHAPTER 3

Chapter 1

1. HOPF BIFURCATION PROBLEM: AN ANALYTICAL APPROACH

1.1. Introduction

Consider the dynamical system governed by the system of ordinary differential equations

$$\dot{x} = F(\eta, x) = F_\eta(x) \tag{1}$$

with $x \in \mathbb{R}^n$ and $\eta \in \mathbb{R}^k$. Suppose, that F is smooth and dependent upon parameter η. Let, for η_0, system (1) possesses the constant solution $x(\eta_0) = x_{0\eta_0}$. Generally, system (1) has n+2 topologically different portraits of the phase flow. They correspond to one nonsingular point and n+1 simple singular points $x_0^{p,q}$, where $p = 0,1,\ldots, n$, $p+q = n$ (see [1]). Numbers p and q are the numbers of the eigenvalues of the Jacobian matrix $DF_x(\eta_0,x_0)$ with p negative and q positive real parts. Phase trajectories near the nonsingular point are similar to the parallel lines. The behaviour of the phase trajectories in the neighborhood of the singular point $x_0^{p,q}$ depend on p and q. For p = n all trajectories aspire to the stable point $x_0^{n,0}$ when $t \to +\infty$, whereas for q = n all the trajectories aspire to the unstable point $x_0^{0,q}$ when $t \to -\infty$. All the other points $x_0^{p,q}$ are saddle points. Two surfaces S_p^+ and S_q^- with codimension p and q, respectively pass through the saddle point. If the trajectory belongs to the surface S_p^+ (S_q^-) the point $x_0^{p,q}$ is the stable one $x_0^{p,0}$ (unstable one $x_0^{0,q}$). The stable (unstable) singular point corresponds to the stable (unstable) equilibrium point of the considered dynamical system. By continuously changing the parameter η the numbers p and q change. This change takes place if one (or a pair) of eigenvalues of the characteristic equation

$\det \| DF_x(x_0) - \lambda I \| = 0$ cross the imaginary axis of the complex plane λ. This means, that a bifurcation from an equilibrium state appears either when a simple zero eigenvalue or a simple pair of pure imaginary eigenvalues passes through the imaginary axis. The latter case known as a Hopf bifurcation is investigated in this chapter.

The Hopf bifurcation is up to now widely discribed in literature. The extensive list of references connecting with this question is for instance given by Marsden and McCracken [2], Iooss and Joseph [3], Hassard, Kazarinoff and Wan [4]. The aim of this chapter is to present applications of approximate analytical methods to obtain the new bifurcated solutions (i. e. perturbation technique and harmonic balance method). Mentioned approximate analytical methods are well documented (see for instance [5-9]). In this chapter three different methods are presented to solve the Hopf bifurcation problem. In the first considered example $\eta \in \mathbb{R}^1$ and one pair of purely imaginary eigenvalues cross the imaginary axis. The use of the center manifold theorem [10] allows the approximate solution, which exactness is improved by the method of successive approximations, to be easily obtained. In the second example with $\eta \in \mathbb{R}^2$ the new periodic solution is sought in the form of a power series in relation to parameter η and another parameter connected with the amplitude of oscillations. The third example presents the method to obtain quasiperiodic solutions which appear after two pairs of purely imaginary eigenvalues cross the imaginary axis. In this case two independent perturbation parameters ε_1 and ε_2 (connected with the amplitudes of oscillations) are involved in order to obtain postbifurcation solutions. The Hopf bifurcation problem in autonomous systems is widely investigated by Huseyin and coworkers (see for instance [11-14]).

The investigation of external periodic forcing and parametric periodic perturbations on the Hopf bifurcating solutions are more complicated when compared with the autonomous systems [15-18] and is also considered here.

All the results of this chapter are based on the results given in [19-24] (see also [25,26]).

1.2. One Parameter Hopf Bifurcation

Consider the differential equation (1) where $\eta \in \mathbb{R}^1$. Suppose, that the characteristic equation which corresponds to the Jacobian matrix $DF_\eta\ (x_0, \eta)$ has the form

$$(\sigma - \sigma_1(\eta))(\sigma - \sigma_2(\eta))\ P(\eta, \sigma) = 0, \tag{2}$$

where

$$\begin{aligned} &\sigma_{1,2} = \xi(\eta) \pm i\,\omega(\eta), \\ &\omega(\eta_c) \neq 0, \\ &\xi(\eta_c) = 0, \\ &\xi_\eta(\eta_c) \neq 0, \end{aligned} \tag{3}$$

and η_c is the critical value of parameter η. Assume that equation (1) can be presented in the form

$$\dot{u} = K_u(\eta)u + K_v(\eta)v + \tilde{K}(\eta, u, v),$$

$$\dot{v} = S_u(\eta)u + S_v(\eta)v + \tilde{S}(\eta, u, v), \tag{4}$$

where $u \in \mathbb{R}^2$, $v \in \mathbb{R}^{n-2}$, $\eta \in \mathbb{R}^1$, $x = \mathrm{col}[u,v]$ and matrixes $K_u(\eta)$, $K_v(\eta)$, $S_u(\eta)$, $S_v(\eta)$ are the linear parts of expansion of $F(x,\eta)$ into the Taylor series around the equilibrium point x_0 ($x_0 = 0$ is taken into further consideration). Equation $P(\eta, \sigma) = 0$ has roots with negative real parts, for η in the neighbourhood of η_c, and σ_1 and σ_2 are eigenvalues of the two dimensional matrix $K_u(\eta_c)$. The matrix K_u, known also as the critical matrix, determines the Hopf bifurcation.

From the center manifold theorem, in the vicinity of the equilibrium path $x_0 = 0$, there exists a function, $v = f(u)$

which has the property $f_u = 0$†. For the first approximation, one can assume that

$$v = f(u) = 0. \tag{5}$$

Taking (5) in (4) into account one can obtain

$$\dot{u} = K_u(\eta)u + \tilde{K}(\eta, v), \tag{6}$$

where $\tilde{K}(\eta,u) = \tilde{K}(\eta,u,0)$. In the neighbourhood of the critical point

$$K_u(\eta) = K_u(\eta_c) + K_{u\eta}(\eta - \eta_c) + \frac{1}{2}K_{u\eta\eta}(\eta - \eta_c)^2 + \dots \tag{7}$$

Let $u(\eta) = 0$ be the solution of (6) for $\eta < \eta_c$ and at the point $\eta = \eta_c$ the periodic solution $u(t,\varepsilon) = u(t + T,\varepsilon)$ of the period T bifurcates from equilibrium. This solution depends on the formally assumed perturbation parameter connected with the amplitude of oscillations. Then dimensionless time $\tau = \omega t$ is introduced and from (6) one obtains

$$\omega(\varepsilon)\, u_\tau(\tau,\varepsilon) = K'(\eta,\tau,\varepsilon), \tag{8}$$

where $K'(\eta,\tau,\varepsilon) = K_u(\eta)u(\tau,\varepsilon) + \tilde{K}(\eta,u(\tau,\varepsilon))$ and $u(\tau,\varepsilon) = u(\tau + 2\pi,\varepsilon)$ and $\omega(\varepsilon)$ is the unknown frequency. Periodic solutions $u(\tau,\varepsilon)$ of equation (8) are searched in the form of a particular Fourier series, which coefficients depend on the parameter ε

$$u_i(\tau,\varepsilon) = \sum_{k=0}^{K} (p_{ik}(\varepsilon)\cos k\tau + r_{ik}(\varepsilon)\sin k\tau). \tag{9}$$

Moreover, $p_{ik}(\varepsilon), r_{ik}(\varepsilon)$, $\eta(\varepsilon)$ and $\omega(\varepsilon)$ are developed into the power series of the parameter ε

† In the notation used in this chapter this character denotes the differentiation.

$$p_{ik}(\varepsilon) = p_{ik0} + p'_{ik}\varepsilon + \frac{1}{2} p''_{ik}\varepsilon^2 + \dots ,$$

$$r_{ik}(\varepsilon) = r_{ik0} + r'_{ik}\varepsilon + \frac{1}{2} r_{ik}\varepsilon^2 + \dots ,$$

$$\eta(\varepsilon) = \eta_c + \eta'\varepsilon + \frac{1}{2}\eta''\varepsilon^2 + \dots , \tag{10}$$

$$\omega(\varepsilon) = \omega_c + \omega'\varepsilon + \frac{1}{2}\omega''\varepsilon^2 + \dots ,$$

where in the critical point $p_{ik0} = r_{ik0} = 0$. The solution $u_i(\tau,\varepsilon)$ is searched in the form

$$u_i(\tau,\varepsilon) = u'_i(\varepsilon) \quad + \frac{1}{2}u''_i(\tau)\varepsilon^2 + \dots , \tag{11}$$

where

$$u_i^{(*)}(\tau) = \sum_{k=1}^{K} (p_{ik}^{(*)} \cos k\tau + r_{ik}^{(*)} \sin k\tau). \tag{12}$$

Because the system (8) is autonomous we set a $r_{11}(\varepsilon) = 0$. Now one can act in one of two ways. Either introduce equations (9) to (12) into (8) and then by comparing the terms with the same powers of ε obtain a linear differential equations set , or obtain these equations by successive differentiation of (8). Employing the latter possibility one obtains

$$\omega' u_{i\tau} + \omega_c u'_{i\tau} = \frac{\partial K'_i}{\partial u_1}\frac{\partial u_1}{\partial \varepsilon} + \frac{\partial K'_i}{\partial u_2}\frac{\partial u_2}{\partial \varepsilon} + \frac{\partial K'_i}{\partial \eta}\eta' \quad ,$$

$$\omega'' u_{i\tau} + 2\omega' u_{i\tau} + \omega_c u''_{i\tau} = \frac{\partial^2 K'_i}{\partial (u_1)^2}\left(\frac{\partial u_1}{\partial \varepsilon}\right)^2 + \frac{\partial^2 K'_i}{\partial (u_2)^2}\left(\frac{\partial u_2}{\partial \varepsilon}\right)^2$$

$$+ 2 \frac{\partial^2 K_i'}{\partial u_1 \partial u_2} \frac{\partial u_1}{\partial \varepsilon} \frac{\partial u_2}{\partial \varepsilon} + 2 \frac{\partial^2 K_i'}{\partial u_1 \partial \eta} \frac{\partial u_1}{\partial \varepsilon} \eta' + 2 \frac{\partial^2 K_i'}{\partial u_2 \partial \eta} \frac{\partial u_2}{\partial \varepsilon} \eta'$$

$$+ \frac{\partial^2 K_i'}{\partial (u_2)^2} \left(\frac{\partial u_2}{\partial \varepsilon}\right)^2 + \frac{\partial K_i'}{\partial \eta} \eta'' + \frac{\partial K_i'}{\partial u_1} \frac{\partial^2 u_1}{\partial \varepsilon^2} + \frac{\partial K_i'}{\partial u_2} \frac{\partial^2 u_2}{\partial \varepsilon^2}$$

$$+ \frac{\partial^2 K_i'}{\partial \eta^2} (\eta')^2,$$

$$\omega''' u_{i\tau} + 3\omega'' u_{i\tau} + 3\omega' u_{i\tau}'' + \omega_c u_{i\tau}''' = \frac{\partial K_i'}{\partial u_1} \frac{\partial^3 u_1}{\partial \varepsilon^3} + \frac{\partial K_i'}{\partial u_2}$$

$$\cdot \frac{\partial^3 u_2}{\partial \varepsilon^3} + 3 \frac{\partial^2 K_i'}{\partial u_1 \partial \eta} \frac{\partial u_1}{\partial \varepsilon} \eta'' + 3 \frac{\partial^2 K_i'}{\partial u_2 \partial \eta} \frac{\partial u_2}{\partial \varepsilon} \eta'' + 3 \frac{\partial^2 K_i'}{\partial u_1 \partial \eta} \frac{\partial^2 u_1}{\partial \varepsilon^2} \eta'$$

$$+ 3 \frac{\partial^2 K_i'}{\partial u_2 \partial \eta} \frac{\partial^2 u_2}{\partial \varepsilon^2} \eta' + 6 \frac{\partial^3 K_i'}{\partial u_1 \partial u_2 \partial \eta} \frac{\partial u_1}{\partial \varepsilon} \frac{\partial u_2}{\partial \varepsilon} \eta' + 3 \frac{\partial^3 K_i'}{\partial u_1 \partial \eta^2}$$

$$\cdot \frac{\partial u_1}{\partial \varepsilon} (\eta')^2 + 3 \frac{\partial^3 K_i'}{\partial u_2 \partial \eta^2} \frac{\partial u_2}{\partial \varepsilon} (\eta')^2 + 3 \frac{\partial^2 K_i'}{\partial u_1 \partial u_2} \frac{\partial^2 u_1}{\partial \varepsilon^2} \frac{\partial u_2}{\partial \varepsilon}$$

$$+ 3 \frac{\partial^2 K_i'}{\partial u_1 \partial u_2} \frac{\partial^2 u_2}{\partial \varepsilon^2} \frac{\partial u_1}{\partial \varepsilon} + 3 \frac{\partial^2 K_i'}{\partial (u_1)^2} \frac{\partial^2 u_1}{\partial \varepsilon^2} \frac{\partial u_1}{\partial \varepsilon} + 3 \frac{\partial^2 K_i}{\partial (u_2)^2}$$

$$\cdot \frac{\partial^2 u_2}{\partial \varepsilon^2} \frac{\partial u_2}{\partial \varepsilon} + \frac{\partial^3 K_i'}{\partial (u_1)^3} \left(\frac{\partial u_1}{\partial \varepsilon}\right)^3 + \frac{\partial^3 K_i'}{\partial (u_2)^2} \left(\frac{\partial u_2}{\partial \varepsilon}\right)^3 + 3 \frac{\partial^3 K_i'}{\partial (u_1)^2 \partial u_2}$$

$$\cdot\left(\frac{\partial u_1}{\partial \varepsilon}\right)^2 \frac{\partial u_2}{\partial \varepsilon} + 3\,\frac{\partial^3 K_i'}{\partial u_1\, \partial (u_2)^2} \left(\frac{\partial u_2}{\partial \varepsilon}\right)^2 \frac{\partial u_1}{\partial \varepsilon}\,, \tag{13}$$

$$\vdots \qquad\qquad\qquad\qquad \vdots$$

Solving equations (13) successively, we obtain (9) and (10). Solutions $u \in \mathbb{R}^2$ obtained in this way are denoted as u_b and put in the second subsystem of equations (4)

$$\omega_c v_\tau = S_u(\eta) u_b + S_v(\eta) v + \tilde{S}(\eta, u_b, v), \tag{14}$$

where ω, η and u_b are known. In order to obtain v, this system of equations can be solved by a perturbation method.

1.3. Biparameter Hopf Bifurcation

Consider the case when $\eta = (\mu, \delta)$. Taking $\tau = \omega t$ into account one obtains from (1)

$$\omega x_\tau = F(x, \eta). \tag{15}$$

Periodical bifurcation solutions are searched in the form

$$x_i(\tau; \mu, \varepsilon) = x_i^{\varepsilon}(\tau) + \frac{1}{2} x_i^{\varepsilon^2}(\tau)\varepsilon^2 + \frac{1}{6} x_i^{\varepsilon^3}(\tau)\varepsilon^3 + x_i^{\mu\varepsilon}(\tau)$$

$$\cdot \mu\varepsilon + \frac{1}{2} x_i^{\mu^2\varepsilon}(\tau)\varepsilon\mu^2 + \frac{1}{2} x_i^{\varepsilon^2\mu}(\tau)\varepsilon^2\mu + \ldots \tag{16}$$

where ε is a certain perturbation parameter connected with the amplitude of oscillations. Every component of the series (16) is searched in the form

$$x_i^{(*)}(\tau) = \sum_{m=0}^{M} (p_{im}^{(*)} \cos m\tau + r_{im}^{(*)} \sin m\tau). \tag{17}$$

Coefficients p_{im}, r_{im}, frequency ω and δ are searched in the form of the following series

$$p_{im}(\varepsilon,\mu) = p_{imc} + p_{im}^{\varepsilon}\varepsilon + \frac{1}{2}p_{im}^{\varepsilon^2}\varepsilon^2 + \frac{1}{6}p_{im}^{\varepsilon^3}\varepsilon^3 + p_{im}^{\mu\varepsilon}\mu\varepsilon$$

$$+ \frac{1}{2}p_{im}^{\mu^2\varepsilon}\mu^2\varepsilon + \frac{1}{2}p_{im}^{\mu\varepsilon^2}\mu\varepsilon^2 + \ldots,$$

$$r_{im}(\varepsilon,\mu) = r_{imc} + r_{im}\varepsilon + \frac{1}{2}r_{im}^{\varepsilon^2}\varepsilon^2 + \frac{1}{6}r_{im}^{\varepsilon^3}\varepsilon^3 + r_{im}^{\mu\varepsilon}\mu\varepsilon$$

$$+ \frac{1}{2}r_{im}^{\mu^2\varepsilon}\mu^2\varepsilon + \frac{1}{2}p_{im}^{\mu\varepsilon^2}\mu\varepsilon^2 + \ldots, \tag{18}$$

$$\omega(\mu,\varepsilon) = \omega_c + \omega_\varepsilon\varepsilon + \frac{1}{2}\omega_{\varepsilon\varepsilon}\varepsilon^2 + \frac{1}{6}\omega_{\varepsilon\varepsilon\varepsilon}\varepsilon^3 + \omega_{\mu\varepsilon}\mu\varepsilon$$

$$+ \frac{1}{2}\omega_{\mu\varepsilon^2}\mu\varepsilon^2 + \frac{1}{2}\omega_{\varepsilon\mu^2}\varepsilon\mu^2 + \omega_\mu\mu + \frac{1}{2}\omega_{\mu\mu}\mu^2 + \ldots,$$

$$\delta(\mu,\varepsilon) = \delta_c + \delta_\varepsilon\varepsilon + \frac{1}{2}\delta_{\varepsilon\varepsilon}\varepsilon^2 + \frac{1}{2}\delta_{\varepsilon\varepsilon\varepsilon}\varepsilon^3 + \delta_{\mu\varepsilon}\mu\varepsilon$$

$$+ \frac{1}{2}\delta_{\mu\varepsilon}2\mu\varepsilon^2 + \frac{1}{2}\delta_{\varepsilon\mu}2\,\varepsilon\mu^2 + \mu\delta_\mu + \frac{1}{2}\delta_{\mu\mu}\mu^2 + \ldots$$

The subsequent differentiation of (15) in relation to ε and μ gives

$$\omega_\varepsilon x_{i\tau} + \omega_c x^\varepsilon_{i\tau} = F_{ij} x^\varepsilon_j,$$

$$\omega_{\varepsilon\varepsilon} x_{i\tau} + 2\omega_\varepsilon x^\varepsilon_{i\tau} + \omega_c x^{\varepsilon^2}_{i\tau} = F_{ijk} x^\varepsilon_j x^\varepsilon_k + 2F_{ij\delta} x^\varepsilon_j \delta_\varepsilon + F_{ij} x^{\varepsilon^2}_j,$$

$$\omega_{\varepsilon\mu} x_{i\tau} + \omega_\mu x^\varepsilon_{i\tau} + \omega_c x^{\mu\varepsilon}_{i\tau} = F_{ij} x^{\varepsilon\mu}_j$$

$$+ F_{ij\delta} \delta_\mu x^\varepsilon_j + F_{ij\mu} x^\varepsilon_j + F_{i\delta\delta} \delta_\varepsilon \delta_\mu + F_{i\delta\mu} \delta_\varepsilon ,$$

$$\omega_{\varepsilon\varepsilon\varepsilon} x_{i\tau} + 3\omega_{\varepsilon\varepsilon} x^\varepsilon_{i\tau} + 3\omega_\varepsilon x^{\varepsilon^2}_{i\tau} + \omega_c x^{\varepsilon^3}_{i\tau} = F_{ijkl} x^\varepsilon_j x^\varepsilon_k x^\varepsilon_l$$

$$+ 3F_{ijk\delta} x^\varepsilon_j x^\varepsilon_k \delta_\varepsilon + 3F_{ijk} x^{\varepsilon^2}_j x^\varepsilon_k + 3F_{ij\delta\delta} x^\varepsilon_j (\delta_\varepsilon)^2$$

$$+ 3F_{ij\delta} x^{\varepsilon^2}_j \delta_\varepsilon + 3F_{ij\delta} x^\varepsilon_j \delta_{\varepsilon\varepsilon} + F_{ij} x^{\varepsilon^3}_j, \qquad (19)$$

$$\omega_{\varepsilon\mu^2} x_{i\tau} + \omega_{\mu\mu} x^\varepsilon_{i\tau} + 2\omega_\mu x^\varepsilon_{i\tau} + \omega_c x^{\varepsilon\mu^2}_{i\tau} = F_{ijk} x^\varepsilon_j x^{\mu\varepsilon}_k$$

$$+ 2F_{ij\mu} x^{\varepsilon\mu}_j + 2F_{ij\delta} \delta_\mu x^{\varepsilon\mu}_j + F_{ij} x^{\varepsilon\mu^2}_j + 2F_{ij\delta\mu} \delta_\mu x^\varepsilon_j$$

$$+ F_{ij\delta\delta} (\delta_\mu)^2 x^\varepsilon_j + F_{ij} \delta_{\mu\mu} x^\varepsilon_j + F_{ij\mu\mu} x^\varepsilon_j + F_{ij\delta} \delta_\varepsilon x^{\mu\varepsilon}_j$$

$$+ F_{i\delta\delta\delta} \delta_\varepsilon (\delta_\mu)^2 + 2F_{i\delta\delta\mu} \delta_\varepsilon \delta_\mu + F_{i\delta\delta} \delta_{\varepsilon\mu} \delta_\mu + F_{i\delta\delta} \delta_\varepsilon \delta_{\mu\mu}$$

$$+ F_{i\delta\mu\mu} \delta_\varepsilon + F_{i\delta\mu} \delta_{\varepsilon\mu},$$

$$\omega_{\varepsilon^2\mu} x_{i\tau} + 2\omega_{\varepsilon\mu} x_{i\tau}^{\varepsilon} + 2\omega_{\varepsilon} x_{i\tau}^{\varepsilon\mu} + \omega_{\mu} x_{i\tau}^{\varepsilon^2} + \omega_c x_i^{\varepsilon^2\mu} =$$

$$F_{ijk\mu} x_j^{\varepsilon} x_k^{\varepsilon} + F_{ijk} x_j^{\varepsilon\mu} x_k^{\varepsilon} + F_{ijk} x_j^{\varepsilon} x_k^{\varepsilon\mu} + F_{ijk\delta} \delta_{\mu} x_j^{\varepsilon} x_k^{\varepsilon}$$

$$+ 2F_{ij} \delta_{\mu} x_j^{\varepsilon} \delta_{\varepsilon} + 2F_{ij\delta} x_j^{\varepsilon\mu} \delta_{\varepsilon} + 2F_{ij\delta} x_j^{\varepsilon} \delta_{\mu\varepsilon} + 2F_{ij\delta\delta} (\delta_{\mu})^2 x_j^{\varepsilon} \delta_{\varepsilon}$$

$$+ F_{ij\mu} x_j^{\varepsilon^2} + F_{ij} x_j^{\varepsilon^2\mu} + F_{ij\delta} \delta_{\mu} x_j^{\varepsilon^2},$$

where

$$\omega_{\varepsilon\varepsilon\varepsilon} = \frac{\partial^3 \omega}{\partial \varepsilon^3}; \quad \omega_{\varepsilon^2\mu} = \frac{\partial^3 \omega}{\partial \varepsilon^2 \partial \mu}, \ \ldots, \ F_{ijk} = \frac{\partial^2 F_i}{\partial x_j \partial x_k},$$

$$F_{ijk\delta} = \frac{\partial^3 F_i}{\partial x_j \partial x_k \partial \delta}, \ \ldots$$

Obtained differential perturbation equations (19) are solved by means of the harmonic balance method, having taken into account relations (18) and assumed $r_{11} = 0$.

1.4. Bifurcation into Quasiperiodic Torus

Consider now equations (1) where $\eta \in \mathbb{R}^2$ and where the Jacobian matrix has a block diagonal form

$$DF_x(\eta_c, x_0) = \text{block diag}\left[C_1, C_2, P_1, P_2, \ldots, R_1, R_2, \ldots\right]$$

with real elements. The blocks are given

$$C_1 = \begin{bmatrix} 0 & \omega_{1C} \\ -\omega_{1C} & 0 \end{bmatrix},$$

$$C_2 = \begin{bmatrix} 0 & \omega_{2C} \\ -\omega_{2C} & 0 \end{bmatrix}, \tag{20}$$

$$P_i = \begin{bmatrix} \alpha_i & \omega_i \\ -\omega_i & \alpha_i \end{bmatrix}, \qquad i = 1, 2, \ldots,$$

$$R_j = [\alpha_j], \qquad j = 1, 2, \ldots,$$

while $\alpha_i < 0$ and $\alpha_j < 0$ for $\eta < \eta_c$.
Let $\lambda_{1,2}(\eta_1) = \alpha(\eta_1) \pm i\,\omega_1(\eta_1)$ and $\lambda_{3,4} = \beta(\eta_2) \pm i\,\omega_2(\eta_2)$ be the two pairs of eigenvalues of DF_x that cross the imaginary axis at $\eta_1 = \eta_{1C}$ and $\eta_2 = \eta_{2C}$ with non-zero velocity as η_1 and η_2 are increased. Then for $\eta_1 = \eta_{1C}$ and $\eta_2 = \eta_{2C}$ we have

$$\lambda_{1,2}(\eta_{1C}) = \pm i\,\omega_{1C}\,,$$

$$\lambda_{3,4}(\eta_{2C}) = \pm i\,\omega_{2C} \tag{21}$$

and the real parts of $\lambda_{1,2}$ and $\lambda_{3,4}$ should satisfy

$$\left.\frac{d\alpha(\eta_1)}{d\eta_1}\right|_{\eta_1 = \eta_{1C}} \neq 0,$$

$$\left.\frac{d\beta(\eta_2)}{d\eta_2}\right|_{\eta_2 = \eta_{2C}} \neq 0. \tag{22}$$

We consider the non-resonant case, i. e. when $\frac{\omega_{1C}}{\omega_{2C}} \neq \frac{k}{l}$, where k, l are integer numbers.
Because of the analogy to the Hopf bifurcation this case will be refered to as called double Hopf bifurcation.

The bifurcated solutions of (1) are assumed to be in the parametric form given by

$$x_i = x_i(t, \varepsilon_1, \varepsilon_2). \tag{23}$$

The solutions (23) are represented by a double Fourier series

$$x_i = \sum_{k=0}^{K} p^{s}_{iko}(\varepsilon_1, \varepsilon_2)\sin k\tau_1 + \sum_{k=0}^{K} p^{c}_{iko}(\varepsilon_1, \varepsilon_2)\cos k\tau_1$$

$$+ \sum_{l=0}^{L} p^{c}_{iol}(\varepsilon_1, \varepsilon_2)\cos l\tau_2 + \sum_{l=0}^{L} p^{s}_{iol}(\varepsilon_1, \varepsilon_2)\sin l\tau_2$$

$$+ \frac{1}{2}\sum_{k=1}^{K}\sum_{l=1}^{L} p^{cc}_{ikl}(\varepsilon_1, \varepsilon_2)\,[\cos(k\tau_1 + l\tau_2) + \cos(k\tau_1$$

$$- l\tau_2)\big] + p^{cs}_{ikl}(\varepsilon_1,\varepsilon_2)\big[\sin(k\tau_1 + l\tau_2) + \sin(l\tau_2 - k\tau_1)\big]$$

$$+ p^{sc}_{ikl}(\varepsilon_1,\varepsilon_2)\big[\sin(k\tau_1 + l\tau_2) + \sin(k\tau_1 - l\tau_2)\big] \tag{24}$$

$$+ p^{ss}_{ikl}(\varepsilon_1,\varepsilon_2)\big[\cos(k\tau_1 - l\tau_2) - \cos(k\tau_1 + l\tau_2)\big] .$$

The coefficients $p^{(*)}_{i(\cdot)}(\varepsilon_1,\varepsilon_2)$, $r^{(*)}_{i(\cdot)}(\varepsilon_1,\varepsilon_2)$, the frequencies $\omega_1(\varepsilon_1,\varepsilon_2)$, $\omega_2(\varepsilon_1,\varepsilon_2)$ and the bifurcation parameters $\eta_1(\varepsilon_1,\varepsilon_2)$, $\eta_2(\varepsilon_1,\varepsilon_2)$ are searched in the form

$$p^{(*)}_{i(\cdot)}(\varepsilon_1,\varepsilon_2) = p^{(*)}_{ic(\cdot)} + \varepsilon_1 p^{(*)'}_{i(\cdot)} + \frac{1}{2}\varepsilon_1^2 p^{(*)''}_{i(\cdot)} + \frac{1}{6}\varepsilon_1^3 p^{(*)'''}_{i(\cdot)}$$

$$+ \varepsilon_2 p^{(*)}_{i(\cdot)'} + \frac{1}{2}\varepsilon_2^2 p^{(*)}_{i(\cdot)''} + \frac{1}{6}\varepsilon_2^3 p^{(*)}_{i(\cdot)'''} + \varepsilon_1\varepsilon_2 p^{(*)'}_{i(\cdot)'} + \frac{1}{2}\varepsilon_1^2\varepsilon_2 p^{(*)''}_{i(\cdot)'}$$

$$+ \frac{1}{2}\varepsilon_2^2\varepsilon_1 p^{(*)'}_{i(\cdot)''} + \dots , \tag{25}$$

$$r^{(*)}_{i(\cdot)}(\varepsilon_1,\varepsilon_2) = r^{(*)}_{ic(\cdot)} + \varepsilon_1 r^{(*)'}_{i(\cdot)} + \frac{1}{2}\varepsilon_1^2 r^{(*)''}_{i(\cdot)} + \frac{1}{6}\varepsilon_1^3 r^{(*)}_{i(\cdot)}$$

$$+ \varepsilon_2 r^{(*)}_{i(\cdot)'} + \frac{1}{2}\varepsilon_2^2 r^{(*)}_{i(\cdot)''} + \frac{1}{6}\varepsilon_2^3 r^{(*)}_{i(\cdot)'''} + \varepsilon_1\varepsilon_2 r^{(*)''}_{i(\cdot)'} + \frac{1}{2}\varepsilon_1^2\varepsilon_2 r^{(*)''}_{i(\cdot)'}$$

$$+ \frac{1}{2}\varepsilon_2^2\varepsilon_1 r^{(*)'}_{i(\cdot)''} + \dots ,$$

$$\omega_1 = \frac{d\tau_1}{dt} = \omega_{1c} + \varepsilon_1\omega_1' + \frac{1}{2}\varepsilon_1^2\omega_1'' + \frac{1}{6}\varepsilon_1^3\omega_1''' + \varepsilon_2\omega_{1'}$$

$$+ \frac{1}{2}\varepsilon_2^2\omega_{1''} + \frac{1}{6}\varepsilon_2^3\omega_{1'''} + \varepsilon_1\varepsilon_2\omega'_{1'} + \frac{1}{2}\varepsilon_1^2\varepsilon_2\omega''_{1'} + \frac{1}{2}\varepsilon_1\varepsilon_2^2\omega'_{1''} + \dots ,$$

$$\omega_2 = \frac{d\tau_2}{dt} = \omega_{2c} + \varepsilon_1\omega'_2 + \frac{1}{2}\varepsilon_1^2\omega''_2 + \frac{1}{6}\varepsilon_1^3\omega'''_2 + \varepsilon_2\omega_{2'}$$

$$+ \frac{1}{2}\varepsilon_2^2\omega_{2''} + \frac{1}{6}\varepsilon_2^3\omega_{2'''} + \varepsilon_1\varepsilon_2\omega'_{2'} + \frac{1}{2}\varepsilon_1^2\varepsilon_2\omega''_{2'} + \frac{1}{2}\varepsilon_1\varepsilon_2^2\omega'_{2''} + \dots ,$$

$$\eta_1 = \eta_{1c} + \varepsilon_1\eta'_1 + \frac{1}{2}\varepsilon_1^2\eta''_1 + \frac{1}{6}\varepsilon_1^3\eta'''_1 + \varepsilon_2\eta_{1'} + \frac{1}{2}\varepsilon_2^2\eta_{1''}$$

$$+ \frac{1}{6}\varepsilon_2^3\eta_{1'''} + \varepsilon_1\varepsilon_2\eta'_{1'} + \frac{1}{2}\varepsilon_1^2\varepsilon_2\eta''_{1'} + \frac{1}{2}\varepsilon_1\varepsilon_2^2\eta'_{1''} + \dots ,$$

$$\eta_2 = \eta_{2c} + \varepsilon_1\eta'_2 + \frac{1}{2}\varepsilon_1^2\eta''_2 + \frac{1}{6}\varepsilon_1^3\eta'''_2 + \varepsilon_2\eta_{2'} + \frac{1}{2}\varepsilon_2^2\eta_{2''}$$

$$+ \frac{1}{6}\varepsilon_2^3\eta_{2'''} + \varepsilon_1\varepsilon_2\eta'_{2'} + \frac{1}{2}\varepsilon_1^2\varepsilon_2\eta''_{2'} + \frac{1}{2}\varepsilon_1\varepsilon_2^2\eta''_{2'} + \dots .$$

By introducing (24) and (25) into (1) and equating the expressions of the same powers $\varepsilon_1^k\,\varepsilon_2^l (k,l = 0,1,2,\dots)$ a sequence of linear differential equations are obtained, which are solvable by means of the harmonic balance method. Because the system is autonomous it is possible to obtain the dependencies of unknown frequencies and bifurcation parameters on amplitudes of oscillations in this way. Independent small perturbation parameters are connected with amplitudes of oscillations.

1.5. Hopf Bifurcation in Duffing's Oscillator

1.5.1. A vibrating system

Consider the vibrating mechanical system with one degree of

freedom

$$m\frac{d^2x}{dt^2} - (c_1 - c_2x^2)\frac{dx}{dt} + kx + k_1x^3 = P_0\cos\omega_1 t, \qquad (26)$$

where m is the mass of the vibrating body, c_1 and c_2 are damping factors, k and k_1 - rigidity factors, and P_0 and ω_1 are respectively, amplitude and external excitation frequency.

The eq. (26) assumes the dimensionless form

$$\frac{d^2y}{d\tau^2} - \eta\frac{dy}{d\tau} + \delta y^2\frac{dy}{d\tau} + \omega_c^2 y + \xi y^3 = \lambda\cos\tau, \qquad (27)$$

where

$$\tau = \omega_1 t,\ x = \frac{P_{max}}{m\,\omega_1^2}y,\ \eta = \frac{c_1}{m\omega_1},\ \delta = \frac{c_2P_{max}^2}{m^3\,\omega_1^5},\ \alpha_0^2 = \frac{k}{m},$$

$$\omega_c^2 = \frac{\alpha_0^2}{\omega_1^2},\quad \xi = \frac{k_1P_{max}^2}{m^3\,\omega_1^6},\quad \lambda = \frac{P_0}{P_{max}}. \qquad (28)$$

The eq. (27) is a particular type of the Van der Pol-Duffing equation, and for $\eta = 0$ it circumscribes the vibrations of Duffing's oscillator. It can be presented in the form

$$\frac{dx_1}{d\tau} = x_2 - \frac{\delta x_1^3}{3} + \eta x_1,$$

$$\qquad (29)$$

$$\frac{dx_2}{d\tau} = -\omega_c^2x_1 - \xi x_1^3 + \lambda\cos\tau,$$

where $x_1 = y$. The characteristic equation obtained from (29), in the case of lack of external excitation, has the roots $\sigma_{1,2} = 1/2(\eta\pm(\eta^2-4\,\omega_c^2)^{1/2})$. For $\eta_c = 0$, at the critical point

$\sigma_{1,2} = \pm i\omega_c$ and $d\sigma_{1,2}/d\eta|_c = 1/2$ conditions of occurrence of the Hopf bifurcation are fulfilled.

1.5.2. Non-resonance case

Let $\lambda = \varepsilon\lambda_1$, then the eq. (29) will have the form

$$\frac{dx_1^{(1)}}{d\tau} = x_2^{(1)} - \frac{\delta (x_1^{(1)})^3}{3} + \eta x_1^{(1)},$$

$$\frac{dx_2^{(1)}}{d\tau} = -\omega_c^2 x_1^{(1)} - \xi (x_1^{(1)})^3 + \varepsilon\lambda_1 \cos\tau . \tag{30}$$

The solutions of (30) are searched in the form of series

$$x_i^{(1)} = x_{ic}^{(1)} + \varepsilon x_i^{(1)'} + \frac{1}{2}\varepsilon^2 x_i^{(1)''} + \frac{1}{6}\varepsilon^3 x_i^{(1)'''} + \dots , \tag{31}$$

$i = 1,2.$

$$\eta = \eta_c + \varepsilon\eta' + \frac{1}{2}\varepsilon^2\eta'' + \frac{1}{6}\varepsilon^3\eta''' + \dots , \tag{32}$$

$$\omega = \omega_c + \varepsilon\omega' + \frac{1}{2}\varepsilon^2\omega'' + \frac{1}{6}\varepsilon^3\omega''' + \dots . \tag{33}$$

In the critical point $\eta_c = x_{ic}^{(1)} = 0$. The bifurcation solutions $x_i^{(1)}(\tau, \varepsilon)$ are represented by the Fourier series

$$x_i^{(1)} = \sum_{k=0}^{K} (p_{ik0}^{s}(\varepsilon)\sin k\tau_1 + p_{ik0}^{c}\cos k\tau_1)$$

$$+ \sum_{l=0}^{L} (p_{iol}^{c}(\varepsilon)\cos l\tau + p_{iol}^{s}(\varepsilon)\sin l\tau) + \frac{1}{2}\sum_{l=1}^{L}\sum_{k=1}^{K}\{p_{ikl}^{cc}(\varepsilon)$$

$$[\cos(k\tau_1 + l\tau) + \cos(k\tau_1 - l\tau)] + p_{ikl}^{cs}(\varepsilon)[\sin(k\tau_1$$

$$+ l\tau) + \sin(l\tau - k\tau_1)] + p_{ikl}^{sc}(\varepsilon)[\sin(k\tau_1 + l\tau)$$

$$+ \sin(k\tau_1 - l\tau)] + p_{ikl}^{ss}(\varepsilon)[\cos(k\tau_1 - l\tau) - \cos(k\tau_1$$

$$+ l\tau)]\} \ , \qquad (34)$$

where: $\tau_1 = \omega\tau$,

$$p_{i(\cdot)}^{(*)} = p_{ic(\cdot)}^{(*)} + \varepsilon p_{i(\cdot)}^{(*)'} + \frac{1}{2}\varepsilon^2 p_{i(\cdot)}^{(*)''} + \frac{1}{6}\varepsilon^3 p_{i(\cdot)}^{(*)'''} + \dots \qquad (35)$$

The following parameter-frequency relations and the dependence of the bifurcation parameter on amplitude are obtained

$$\eta = \frac{\delta}{2}\left(\frac{1}{2}A^2 + \frac{\lambda^2}{(\omega_c^2-1)^2}\right) , \qquad (36)$$

$$\omega = \omega_c + \frac{3}{4}\frac{\xi}{\omega_c}\left(\frac{1}{2}A^2 + \frac{\lambda^2}{(\omega_c^2-1)^2}\right) , \qquad (37)$$

where: $A^2 = (\varepsilon p_{110}^{c'})^2 + (\varepsilon p_{110}^{s'})^2$.

1.5.3. Main resonance

Let $\lambda = \varepsilon^3\lambda_2$, then equation (29) gives

$$\frac{dx_1^{(2)}}{d\tau} = x_2^{(2)} - \frac{\delta(x_1^{(2)})^3}{3} + \eta x_1^{(2)} ,$$
$$\frac{dx_2^{(2)}}{d\tau} = -\omega_c^2 x_1^{(2)} - \xi(x_1^{(2)})^3 + \varepsilon^3\lambda_2 \cos\tau . \tag{38}$$

Let us assume that frequency ω_c and the external excitation frequency differ from each other only slightly, i. e. let $\omega_c \cong 1-1/2a'$, where $a' = \varepsilon^2 a$. Now we act analogously to the non-resonance case

$$\eta = \frac{\xi}{4} A^2 - \frac{\lambda}{A} \sin\varphi , \tag{39}$$

$$\omega = \omega_c - \frac{1}{2}\frac{\lambda\cos\varphi}{A} + \frac{3}{8}\xi A^2 , \tag{40}$$

where: $p_{110}^{s'}\varepsilon = A\sin\varphi$; $p_{110}^{c'}\varepsilon = A\cos\varphi$.

1.5.4 Resonance of the n-th order

In this case it is assumed that the following relation exists between the frequency and external excitation frequency: $\omega_c \cong 1/n(1 - 1/2a')$. The equation (29), after taking

$\lambda = \varepsilon\lambda_1$, $\tau = n\tau_2$, gives

$$\frac{dx_1^{(3)}}{d\tau} = x_2^{(3)} - n\frac{\delta}{3}(x_1^{(3)})^3 + n\,\eta\,x_1^{(3)},$$

$$\frac{dx_2^{(3)}}{d\tau} = -x_1^{(3)} + \dot{a}x_1^{(3)} - n^2\xi(x_1^{(3)})^3 + \varepsilon n^2\lambda_1\cos n\tau_2. \tag{41}$$

In this case we obtain

$$\eta = \frac{1}{2}\,\delta\left(\frac{1}{2}A^2 + \frac{\lambda^2 n^4}{(1-n^2)^2}\right), \tag{42}$$

$$\omega = n\,\omega_c + \frac{1}{2}n^2\,\xi\left(\frac{3}{4}A^2 + \frac{3}{2}\,\frac{\lambda^2 n^4}{(1-n^2)^2}\right). \tag{43}$$

1.5.5. Concluding remarks

On the example of the forced Van der Pol-Duffing's oscillator, analytical method of determining the post-critical family of solutions after Hopf bifurcation in nonlinear nonautonomous oscillators with one bifurcation parameter is presented. It is assumed that Hopf's conditions are fulfilled for a certain value of the bifurcation parameter when no forcing exists in the system. The bifurcated solutions are sought in the form of a particular Fourier series. The practical application of the series consists in solving perturbation equations of the first, second and k-th order by means of the harmonic balancing method. The examplary calculations have been reduced to k=3, as an approximate bifurcation solution is determined in principle near the critical point. In the investigated case the amplitude is connected with the small perturbation parameter. The exciting force amplitudes are of the order of the amplitude A, or are smaller.

In the case of no resonance as well as in that of the resonance of n-th order, the parameter ξ has no influence on the dependence of bifurcation parameter, amplitude, amplitude of the exciting force, whereas the parameter δ has no influence on the dependence of amplitude, frequency, amplitude of the exciting force.

1.6. Hopf Bifurction in Nonstationary Nonlinear Systems

1.6.1. Example without external force

Consider a vibrating mechanical system with one degree of freedom — a rotor with unequal moments of inertia of its cross-section, with its mass concentrated in its center. The equation of motion of the system has the form

$$m\ddot{y} + \tfrac{1}{2}(k_1 + k_2 + (k_1 - k_2)\cos 2\omega t)y + k_0 y^3 = 0, \qquad (44)$$

where m is the concentrated mass, k_0 is the non-linear rigidity, k_1, k_2 are the rotor rigidities, and ω is the rotation frequency of the rotor. After a change of variables, a dimensionless form of the equation is obtained,

$$d^2x/d\tau^2 + (\delta^2 + \mu\cos 2\tau)x + \xi x^3 = 0, \qquad (45)$$

where

$$x = y(k_1/k_0)^{-1/2}, \quad \tau = \omega t, \quad \xi = k_1/(m\omega^2),$$

$$\delta^2 = \tfrac{1}{2}(k_1 + k_2)/m\omega^2, \quad \mu = \tfrac{1}{2}(k_1 - k_2)/(k_1 + k_2)\omega^2.$$

Equation (45) can be expressed as a system of two differential equations of the first order,

$$\dot{x}_1 = x_2, \quad \dot{x}_2 = -\delta^2 x_1 - \mu x_1 \cos 2\tau - \xi x_1^3, \tag{46}$$

where $x_1 = y$.

The bifurcation solutions are sought in the form of a series in the perturbation parameter ε connected with the vibration amplitude, of the form

$$x_i = x_i^0 + \varepsilon x_i' + \frac{1}{2}\varepsilon^2 x_i'' + \frac{1}{6}x_i'''\varepsilon^3 + \dots , \tag{47}$$

and the frequency δ^2 and the parameter μ also can be developed into power series in the perturbation parameter,

$$\delta^2 = \delta_c^2 + \varepsilon\delta' + \frac{1}{2}\varepsilon^2\delta'' + \frac{1}{6}\varepsilon^3\delta''' + \dots , \tag{48}$$

$$\mu = \mu_c + \varepsilon\mu' + \frac{1}{2}\varepsilon^2\mu'' + \frac{1}{2}\varepsilon^3\mu''' + \dots , \tag{49}$$

where $\delta_c^2 = n^2$ and $\mu_c = 0$. After substituting equations (47) - (49) into equations (46) and developing $x_i^{(\kappa)}$ into Fourier series, harmonic balancing is performed at ε and $\sin n\tau$, $\cos n\tau$ and the following is obtained:

$$p_{2n0}^{s'} = -np_{1n0}^{c'}, \qquad p_{2n0}^{c'} = np_{1n0}^{s'}. \tag{50}$$

Then

$$x_1^3 = \varepsilon^3((p_{1n0}^{c'})^3(\tfrac{3}{4}\cos n\tau + \tfrac{1}{4}\cos 3n\tau) + \tfrac{3}{4}(p_{1n0}^{c'})^2$$

$$\cdot p_{1n0}^{s'}(\sin 3n\tau + \sin n\tau) + \tfrac{3}{4}(p_{1n0}^{s'})^2 p_{1n0}^{c'}(\cos n\tau - \cos 3n\tau)$$

$$+ (p^{s'}_{1n0})^3(\tfrac{3}{4}\sin n\tau - \tfrac{1}{4}\sin 3n\tau)). \tag{51}$$

After harmonic balancing at ε^3, one obtains, for $n \neq 2$,

$$n^2 p^{c'}_{1n0}\,\delta'' = \tfrac{3}{4}((p^{c'}_{1n0})^3 + (p^{s'}_{1n0})^2 p^{c'}_{1n0}),$$

$$n^2 p^{s'}_{1n0}\,\delta'' = \tfrac{3}{4}((p^{s'}_{1n0})^3 + (p^{c'}_{1n0})^2 p^{s'}_{1n0}). \tag{52}$$

The following is the solution of equations (52):

$$\mu'' = p^{c'}_{1n0} = 0, \qquad \delta''_{(1)} = \tfrac{3}{4}(1/n^2)\,\xi\,(p^{s'}_{1n0})^2; \tag{53}$$

$$\mu'' = p^{s'}_{1n0} = 0, \qquad \delta''_{(2)} = \tfrac{3}{4}(1/n^2)\,\xi\,(p^{c'}_{1n0})^2. \tag{54}$$

On the other hand, for $n = 2$ the following is obtained:

$$p^{c'}_{120} = \delta'' = 0, \qquad \mu'' = -\tfrac{3}{16}\,\xi\,(p^{s'}_{120})^2; \tag{55}$$

$$p^{s'}_{120} = \delta'' = 0, \qquad \mu'' = -\tfrac{3}{16}\,\xi\,(p^{c'}_{120})^2. \tag{56}$$

1.6.2. Example including external force

Let now a harmonic force have an effect on the system considered above. In this case the equation of motion has the form

$$m\ddot{y} + \tfrac{1}{2}(k_1 + k_2 + (k_1 - k_2)\cos 2\omega t)y + k_0 y^3 =$$

$$= P_0 \cos \omega_1 t. \tag{57}$$

Let

$$\tau_* = \omega t, \quad x = y(k_1/k_0)^{-1/2}, \quad \lambda_* = (p_0/m\omega^2)(k_0/k_1)^{1/2},$$

$$\xi_* = k_1/(m\omega^2), \quad \delta_*^2 = (k_1 + k_2)/2m\omega^2, \quad \omega_1/\omega = \omega_*,$$

$$\mu_* = (k_1 - k_2)/2m\omega^2. \qquad (58)$$

Then equation (57) will have the form

$$d^2x/d\tau^2 + (\delta_*^2 + \mu_* \cos 2\tau_*)x + \xi_* x^3 = \lambda_* \cos \omega_* \tau_* . \qquad (59)$$

The system of two differential equations replacing equation (59) is

$$\dot{x}_1 = x_2, \quad \dot{x}_2 = -\delta_*^2 x_1 - \xi_* x_1^3 - \mu_* x_1 \cos 2\tau_* + \lambda_* \cos \omega_* \tau_*,$$

$$(60)$$

where $\lambda_* = \varepsilon \lambda_{1*}$. The Fourier series in this case is double and has the independent variable $n\tau_*$ and $l\omega_*\tau_*$.

Consider first the non-resonance case ($l\omega_* \neq n$), where n, $l \in N$. After harmonic balancing at ε and $\sin n\tau_*$, $\cos n\tau_*$ and $\sin l\omega_*\tau_*$, $\cos l\omega_*\tau_*$ and after solving the algebraic equations set one has

$$p_{2n0}^{s'} = -np_{1n0}^{c'}, \quad p_{2n0}^{c'} = np_{1n0}^{s'}; \qquad (61)$$

$$p_{101}^{s'} = p_{201}^{c'} = 0, \quad p_{101}^{c'} = \lambda_1/(n^2 - l^2\omega_*^2),$$

$$p_{201}^{s'} = -l\lambda_1/(n^2 - l^2\omega_*^2). \qquad (62)$$

Then

$$x_1^3 = (p^{c'}_{1n0})^3(\tfrac{3}{4}\cos n\tau_* + \tfrac{1}{4}\cos 3n\tau_*) + \tfrac{3}{4}(p^{c'}_{1n0})^2$$

$$\cdot p^{s'}_{1n0}(\sin 3n\tau_* + \sin n\tau_*) + \tfrac{3}{4}(p^{s'}_{1n0})^2 p^{c'}_{1n0}$$

$$\cdot(\cos n\tau_* - \cos 3n\tau_*) + (p^{s'}_{1n0})^3(\tfrac{3}{4}\sin n\tau_*$$

$$- \tfrac{1}{4}\sin 3n\tau_*) + \tfrac{3}{2}(p^{c'}_{1n0})^2 p^{c'}_{101}\cos 1\omega_*\tau_* + \tfrac{3}{4}(p^{c'}_{1n0})^2$$

$$\cdot p^{c'}_{101}(\cos(2n + 1\omega_*)\tau_* + \cos(2n - 1\omega_*)\tau_*) \qquad (63)$$

$$+ \tfrac{3}{2}p^{c'}_{1n0}p^{s'}_{1n0}p^{c'}_{101}(\sin(2n + 1\omega_*)\tau_* + \sin(2n - 1\omega_*)\tau_*)$$

$$+ \tfrac{3}{2}(p^{s'}_{1n0})^2 p^{c'}_{101}\cos 1\omega_*\tau_* - \tfrac{3}{4}(p^{s'}_{1n0})^2 p^{c'}_{101}(\cos(2n$$

$$+ 1\omega_*)\tau_* + \cos(2n - 1\omega_*)\tau_*) + \tfrac{3}{2}p^{c'}_{1n0}(p^{c'}_{101})^2\cos n\tau_*$$

$$+ \tfrac{3}{4}p^{c'}_{1n0}(p^{c'}_{101})^2(\cos(n + 21\omega_*)\tau_* + \cos(n - 21\omega_*)\tau_*)$$

$$+ \tfrac{3}{2}p^{s'}_{1n0}(p^{c'}_{101})^2\sin n\tau_* + \tfrac{3}{2}p^{s'}_{1n0}(p^{c'}_{101})^2(\sin(n +$$

$$+ 21\omega_*)\tau_* + \sin(n - 21\omega_*)\tau_*) + (p^{c'}_{101})^3(\tfrac{3}{4}\cos 1\omega_*\tau_*$$

$$+ \tfrac{1}{4}\cos 31\omega_*\tau_*).$$

When comparing the terms at orders ε^3 and $\sin n\tau_*$, $\cos n\tau_*$, as well as at $\sin 1\omega_*\tau_*$ and $\cos 1\omega_*\tau_*$ one can obtain μ'' and δ'' and finally, for this case,

$$\mu = 0,$$

$$\delta^2 = \frac{3}{4}\,\xi\,(-\frac{1}{n^2}(\tfrac{1}{2}(p^{c'}_{1n0}\varepsilon)^2 + \tfrac{1}{2}(p^{s'}_{1n0}\varepsilon)^2) + (p^{c'}_{101}\varepsilon)^2)$$

$$+ 1^{-2}\omega_*^{-2}((p_{1n0}^{c'}\varepsilon)^2 + (p_{1n0}^{s'}\varepsilon)^2 + \tfrac{1}{2}(p_{101}^{c'}\varepsilon)^2)). \qquad (64)$$

Let

$$p_{1n0}^{c'}\varepsilon = A\cos\varphi\ , \qquad p_{1n0}^{s'}\varepsilon = A\sin\varphi\ . \qquad (65)$$

Then the following is obtained from using the expressions in (65):

$$\mu = 0,$$

$$\frac{A^2}{2n^2 1^2\omega_*^2/(1^2\omega_*^2+2n^2)} + \frac{\lambda_*^2}{2(n^2-1^2\omega_*^2)1^2\omega_*^2 n^2/(21^2\omega_*^2+n^2)}$$

$$- \frac{\delta^2}{\frac{3}{4}\xi} = 0. \qquad (66)$$

The second of equations (66) has a geometric representation as a quadric surface — a cone.

The parametric forcing occurs when either $n \neq 1$ and $n \cong \omega_* 1$, $n = 1$ and $n \neq \omega_* 1$, or $n = 1 \cong \omega_* 1$. Consider the last most complex case of resonance. Let the following connections occur:

$$1^2\omega_*^2 - 1 = a', \qquad \lambda_* = \varepsilon^2\lambda_1, \qquad (67)$$

where $a' = \varepsilon^2 a$. Then, by proceeding as in the previous case one obtains

$$\delta_*^2/(\omega_*/2) - \frac{A^2}{(2/3\xi)} = 2(\mu_* - a'). \qquad (68)$$

Equation (68) is the equation of hyperbolic paraboloid.

1.6.3. Concluding remarks

An analysis of the Hopf bifurcation in nonstationary nonlinear systems exemplified by the Mathieu-Duffing oscillator has been presented. Considerations have been limited to determining the parameter-frequency relations and the dependence of the bifurcation parameter on the other parameters of the systems. The method of development into series of the perturbation parameter connected with the vibration amplitude (on the assumption that it is a small quantity), and the method of harmonic balancing have been employed. Cases with and without external force have been considered. For the latter case, with the vibration far from resonance, the vibration amplitude A, the amplitude of the excitation force λ_* and the frequency δ, create a quadric surface, which is a cone. On the other hand, for vibrations near resonance, the frequency δ_*, the vibration amplitude A, and the parametric excitation factor μ_* create a surface of the second degree, which is a hyperbolic paraboloid.

Chapter 2

2. BIFURCATION AND CHAOS: NUMERICAL METHOD BASED ON SOLVING BOUNDARY VALUE PROBLEM

2.1. Introduction

Examples of chaotic behaviour in deterministic nonlinear oscillators show that our knowledge cannot explain all the possible complex dynamics of even simple nonlinear systems [27-35]. One can also expect in complicated dynamical systems something more than strange attractors [36]. It is well known that the qualitative change in the behaviour of the investigated dynamical system with the accompanying change of its one or few parameters is due to bifurcations. The standard classical methods based on solving the initial value problem are not efficient enough to give a general structure of the bifurcations in a parameter space. For instance, they do not allow for the accurate calculation of the bifurcation points and for the calculation of unstable (invisible) attractors.

This chapter concentrates on the systematical approach to trace the behaviour of the system by varying one (or more) chosen parameters in autonomous and nonautonomous systems. The presented approach is based on solving a boundary value problem using the shooting method and develops earlier Brommundt's works [37,38], where the Urabe [39,40] method was used. Similar strategy of calculating branching points and new bifurcating branches of solutions is widely described by Seydel [41-44].

2.2. Gradual and Sudden Transition to Chaos

2.2.1. Method and results

A Duffing-type unsymmetric oscillator governed by the equation

$$\ddot{y} + c\dot{y} + y^3 = q + F\cos\omega t \tag{69}$$

is reconsidered. This oscillator was investigated by Ueda [45]. He presented strange attractors for $\omega = 1$, $F = 0.16$, $c = 0.15$, $q = 0.03$ and $q = 0.045$. These results were obtained using the numerical method based on solving the initial value problem. Also Szemplinska-Stupnicka and Bajkowski [46] examined the behaviour of this oscillator by means of an approximate analytical method. The results obtained by them were then verified by the computer simulation analysis.

Here local bifurcations of periodic orbits in the q parameter line are investigated for the following fixed values of parameters: $c = 0.15$, $\omega = 1.0$, $F = 0.16$. The flow of (69) is strictly contracted, which implies only that the saddle-node or the period doubling local bifurcation exists. Hopf bifurcations are excluded. In order to investigate the k-subharmonic solution, the integration interval of (69) is transformed to $2\pi k$ length. By the use of the shooting method it is then possible to solve the following boundary value problem and to obtain multipliers from equations

$$M_i^{(m)}(y_1, y_2, q) - y_i = 0,$$

$$\chi[DM^{(m)}(y_1, y_2, q), \sigma] = 0, \tag{70}$$

$$y_1 = y,\ y_2 = \dot{y},\ i = 1, 2.$$

Above, $M_i^{(m)}$ denotes the i-th component of the stroboscopic phase portrait (Poincaré map) of (69) with nondimensional time $\tau = k\omega t$. χ is the characteristic polynomial whose eigenvalues σ_i are the sought multipliers of the $2\pi k$ periodic Floquet matrix. Calculations were interrupted at the m-th step, only if the following norm $\| M_i^{(m)} - y_i \| = \sum_1^2 | M_i^{(m)} - y_i |^2 \leqslant 10^{-6}$. The multipliers are either real or complex conjugates. We will trace the movement of those multipliers which accompany the change of q.

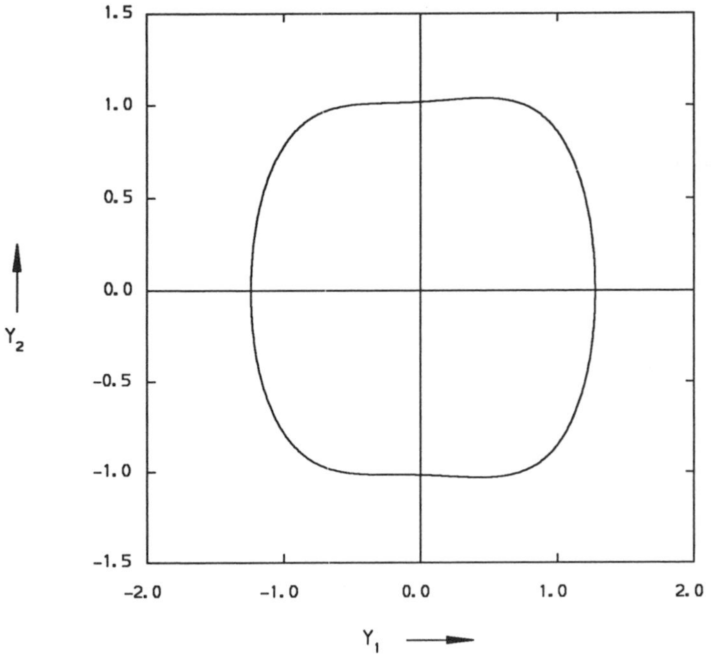

Fig. 1. Phase portrait of the large periodic orbit for $q = 0.03$.

The main resonance solution is presented in Fig. 1. As is seen from Table 1, the increase of q has almost no effect on the behaviour of the multipliers. However, for the considered interval of q, another solution exists (see Table 2). For $q \in (0.042385; 0.03913)$ this is the 1/2 subharmonic solution. There are two possible routes to chaos for this solution (small orbit). The violent burst into a strange chaotic

Table 1. The main resonance solution found using tracing algorithm and corresponding multipliers.

q	y	$\dot{y}$	Multipliers	
			1	2
0.03	1.1406	0.65757	0.21; 0.83	0.21; -0.83
0.035	1.1432	0.66083	0.22; 0.82	0.22; -0.82
0.04	1.1458	0.66408	0.21; 0.83	0.21; -0.83
0.045	1.1483	0.66733	0.21; 0.83	0.21; -0.83
0.05	1.1509	0.67057	0.21; 0.83	0.21; -0.83

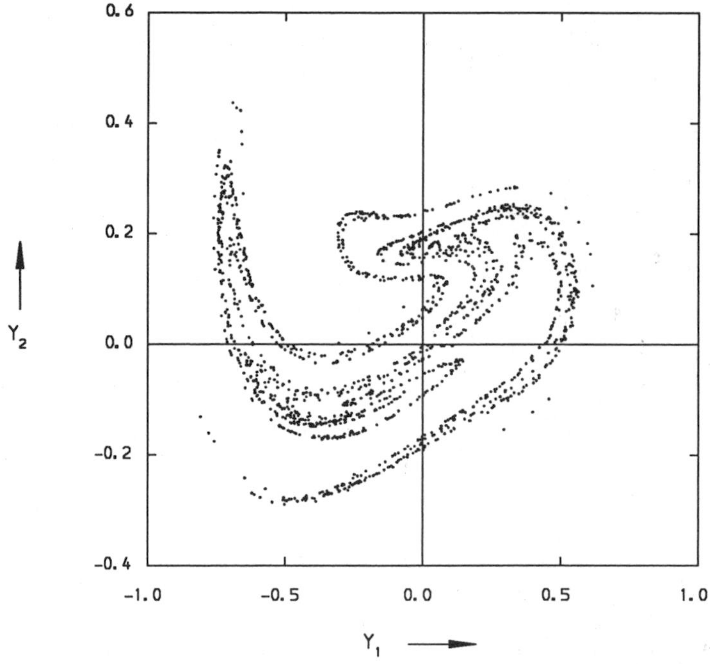

Fig. 2. Strange chaotic attractor obtained for q = 0.043 (one multiplier crosses the unit circle at +1).

Table 2. Bifurcation structure of the small periodic orbit and corresponding eigenvalues-example numerical results. Asterisks are coded bifurcation points.

q	y	$\dot{y}$	Multipliers 1	Multipliers 2	Period
0.042385*	-0.26362	-0.13127	1.0; 0.0	0.54; 0.0	4π
0.042382	-0.26156	-0.13080	0.78; 0.0	0.09; 0.0	
0.042380	-0.26071	-0.13061	0.72; 0.13	0.72;-0.13	
0.042000	-0.23492	-0.12428	0.27; 0.67	0.27;-0.67	
0.041500	-0.22024	-0.12033	0.00; 0.73	0.00;-0.73	
0.040000	-0.19277	-0.11227	-0.54; 0.49	-0.54;-0.49	
0.039300	-0.18314	-0.10923	-0.65; 0.00	-0.83; 0.0	
0.039190*	-0.18173	-0.10877	-0.54; 0.00	-1.0 ; 0.0	
0.039150	-0.18966	-0.10992	0.33; 0.0	0.87; 0.0	8π
0.039110	-0.19487	-0.11066	0.52; 0.10	0.52;-0.10	
0.039000	-0.20149	-0.11156	0.37; 0.38	0.37;-0.38	
0.038300	-0.22495	-0.11436	-0.70; 0.0	-0.40; 0.0	
0.03822*	-0.226810	-0.11455	-1.0 ; 0.0	-0.27; 0.0	
0.03815	-0.235540	-0.11602	-0.01; 0.24	-0.01;-0.24	16π
0.03808	-0.214580	-0.11210	-0.28; 0.32	-0.28;-0.32	
0.03803	-0.21257	-0.11165	-0.23; 0.0	-0.88; 0.0	
0.038029*	-0.21253	-0.11164	-0.08; 0.0	-1.0 ; 0.0	

attractor appears when q is raised beyond 0.042385. For this value of q the multiplier crosses the unit circle of the complex plane at +1 and suddenly another complicated structure of the phase flow is born (Fig. 2). With decreasing q, three steps in the subharmonic scenario were observed. Decreasing q further, the 32π period orbit is extremely sensitive to change. Very small changes of q also cause very large changes in the multipliers, which move inside and outside the unit circle of the complex plane in an unpredictable way. This finally leads to chaos (Fig. 3).

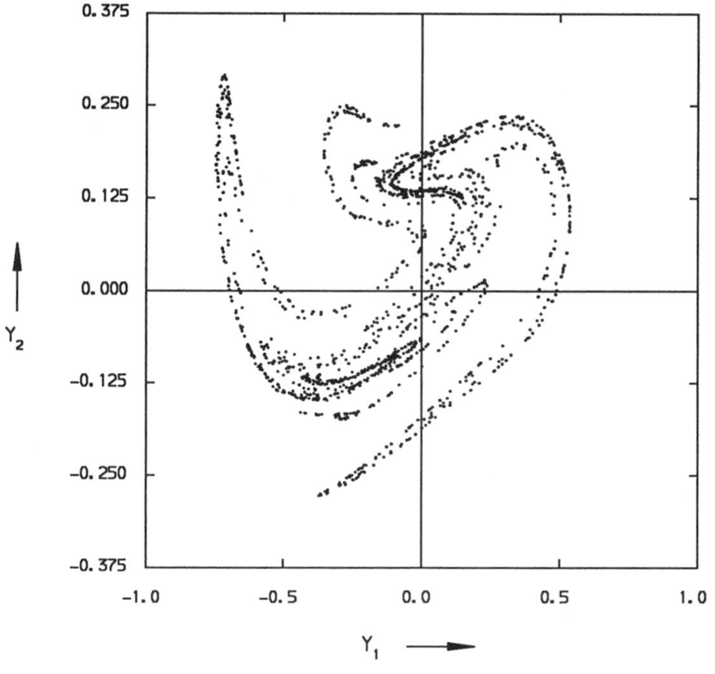

Fig. 3. The strange chaotic attractor obtained after the gradual transition via period doubling bifurcation for q = 0.037.

2.2.2. Conclusions

Two scenarios leading to chaos in the Duffing-type oscillator are investigated with respect to static load by solving the boundary value problem. For the considered interval of q there exist two independent (main resonance and 1/2 subharmonic) solutions. The 1/2 resonance bursts suddenly into chaos if, with increase of the dimensionless static load q, one of the multipliers passes through the unit circle at +1. This possibility is also shown by Neimark [1].

Increasing q draws successive subharmonic bifurcations. However, starting with 32π period the system is extremely sensitive to very small changes of q, making further visualisation very hard. Many bifurcations which have appeared in a rather stochastic way, have been observed after chaos is finally reached.

To summarize the results presented, it is shown how in two different ways the 1/2 subharmonic solution (small orbit) can reach chaos with changing q. With increase of q, the strange attractor appeared suddenly and the previous periodic orbit disappeared. Contrary to this scenario, another accompanying decrease in q led gradually to chaotic motion. Additionally, all of the subharmonic solutions obtained continued to exist since only their stability had changed.

The main resonance (large orbit) has no influence on the behaviour of the small orbit and is stable in the interval of q considered. The transition between the main resonance and the strange chaotic attractor is only possible with a sudden change in initial conditions - "jump" phenomenon.

2.3. Three Different Routes Leading to Chaos

2.3.1 . Period doubling bifurcation

Consider the oscillator governed by the equation

$$\ddot{y} + c\dot{y} - 0.5(1 - y^2)y = P\cos\omega t. \qquad (71)$$

This simple equation models the oscillations of a buckled beam and was examined in [31] . Equation (71) is transformed to

$$y'' + czy' - 0.5z^2 (1 - y^2)y = Pz^2\cos\tau , \qquad (72)$$

where $' = d/d\tau$, $\tau = \omega t$, $z = 1/\omega$, in order to use the calculation technique. It is well known [47] , that if $P = 0$ the system possesses three equilibria, two sinks and a saddle. For the small value of P, the two sinks at ± 1 become small periodic orbits of period $2\pi z$ and a saddle point becomes a saddle type orbit. By increasing the amplitude of the exciting force the system has three periodic orbits. Two small orbits around ± 1 and one large orbit. For certain parameter values the left small periodic orbit and the large one become unstable. The question arises, if it is possible to find parameters for which the last small periodic orbit becomes unstable. Then one can expect chaotic orbits to appear.

The oscillator (71) was recently reconsidered by Szemplinska-Stupnicka [48]. In her analytical approach the refined approximate criterion for chaos has been developed. This criterion was based on the investigation of stability of the small orbit.

During our numerical calculations the following parameters are fixed: $c = 0.1$, $z = 1.25$. P is the bifurcation parameter and the behaviour of the system is traced by use of linear prediction. The results of numerical calculations are included in Table 3, where the corresponding multipliers are also given. For the marked values (*) of P, phase portraits of the system are presented in Fig. 4.

Table 3. The main and successive bifurcation solutions for the oscillator (71).

P	y	$\dot{y}$	Eigenvalues 1	Eigenvalues 2	Period
0.07 *	0.33815	0.19275	-0.63;-0.25	-0.63; 0.25	T=2π
0.08 ▪	0.32750	0.18540	-0.45; 0.0	-1. ; 0.0	
0.08020	0.36025	0.19731	0.15; 0.0	0.84; 0.0	T=4π
0.0804	0.36909	0.20025	0.29; 0.0	0.72; 0.0	
0.0806	0.37650	0.20265	0.36; 0.0	0.57; 0.0	
0.0808	0.38299	0.20468	0.43; 0.15	0.43;-0.15	
0.0810	0.38880	0.20645	0.39; 0.23	0.39;-0.23	
0.0815	0.40149	0.21015	0.30; 0.35	0.30;-0.35	
0.0820	0.41242	0.21314	0.20; 0.41	0.20;-0.41	
0.083 *	0.43110	0.21783	0.01; 0.4	0.01;-0.4	
0.085	0.46134	0.22426	-0.39; 0.24	-0.39;-0.24	
0.086	0.47432	0.22655	-0.96; 0.0	-0.22; 0.0	
0.0863 ▪	0.48236	0.22728	-1. ; 0.0	-0.28; 0.0	
0.08625	0.49096	0.22956	0.64; 0.0	0.07; 0.0	T=8π
0.08630	0.49313	0.22991	0.60; 0.0	0.01; 0.0	
0.0864 *	0.49683	0.23047	0.2 ; 0.06	0.2 ;-0.06	
0.08645	0.49851	0.23070	0.15; 0.15	0.15;-0.15	
0.08650	0.50008	0.23092	0.10; 0.19	0.10;-0.19	
0.08660	0.50292	0.23130	-0.01; 0.21	-0.01;-0.21	
0.0870	0.51228	0.23240	-0.81; 0.0	-0.05; 0.0	
0.08708 ▪	0.51430	0.23261	-1. ; 0.0	-0.03; 0.0	

(Contd.)

Table 3 (contd.)

P	y	$\dot{y}$	Eigenvalues 1	Eigenvalues 2	Period
0.087105	0.51549	0.23249	0.76; 0.0	0.01; 0.0	
0.087108	0.51561	0.23248	0.72; 0.0	0.03; 0.0	
0.08712	0.51598	0.23247	0.63; 0.0	0.0 ; 0.0	
0.08713	0.51627	0.23247	0.0 ; 0.0	0.56; 0.0	
0.08714	0.51656	0.23245	-0.01; 0.0	0.45; 0.0	
0.08715	0.51682	0.23245	-0.01; 0.0	0.35; 0.0	
0.08716	0.51708	0.23244	-0.08; 0.0	0.32; 0.0	
0.08717	0.51731	0.23244	-0.03; 0.0	0.19; 0.0	$T=16\pi$
0.08718 *	0.51755	0.23243	-0.05; 0.0	0.1 ; 0.0	
0.08720	0.51800	0.23243	-0.18; 0.0	0.04; 0.0	
0.08723	0.51863	0.23242	-0.91; 0.0	0.25; 0.0	
0.08724	0.51883	0.23241	-0.58; 0.0	0.02; 0.0	
0.08727	0.51941	0.23241	-0.90; 0.0	0.01; 0.0	
0.08730	0.51997	0.23240	-0.38; 0.40	-0.38;-0.40	
0.087316 •	0.52028	0.23240	-1.0 ; 0.0	0.02; 0.0	
0.08732 *	0.52125	0.23263	-0.37; 0.25	-0.37;-0.25	
0.087323	0.52133	0.23264	-0.31; 0.0	-0.63; 0.0	
0.087325	0.52138	0.23264	-0.26; 0.0	-0.76; 0.0	$T=32\pi$
0.087328	0.52146	0.23265	-0.23; 0.0	-0.96; 0.0	
0.087329 •	0.52149	0.23266	-0.22; 0.0	-1.0 ; 0.0	

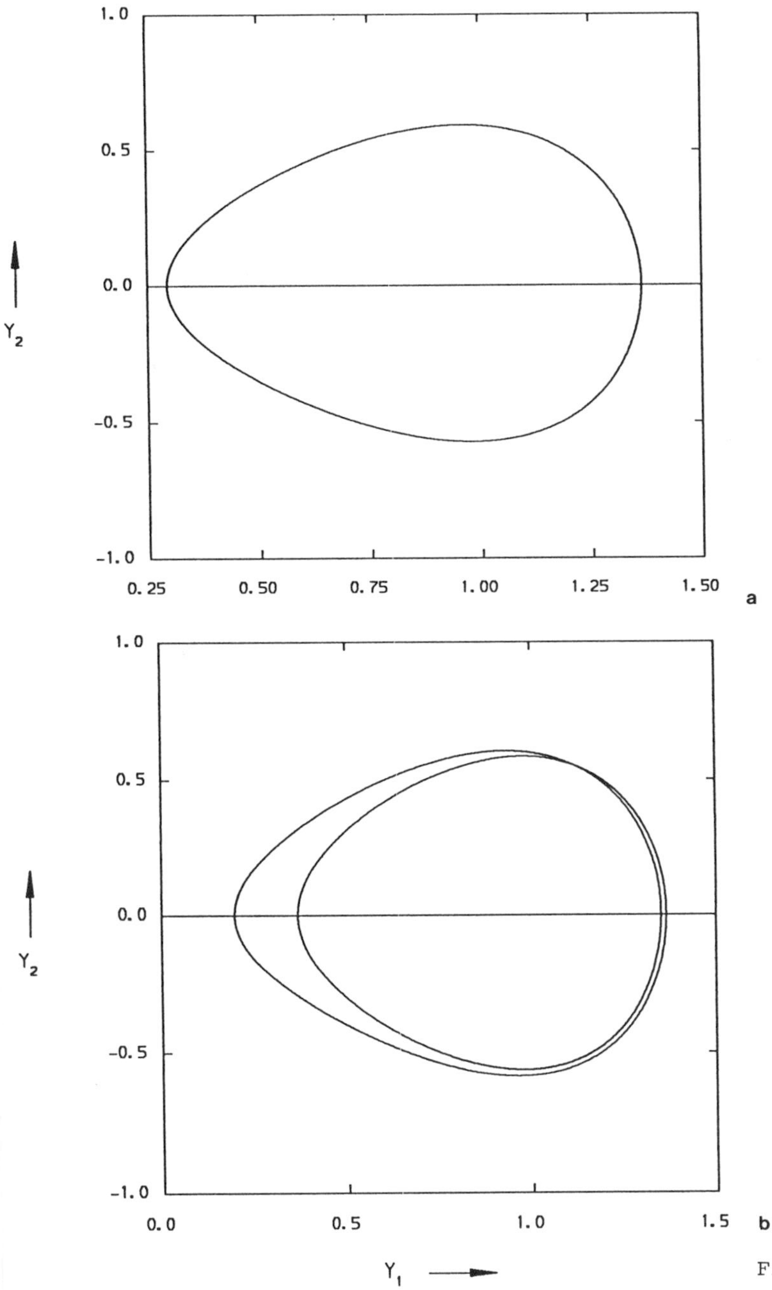
1.0
0.5
0.0
-0.5
-1.0
0.25
0.50
0.75
1.00
1.25
1.50
a
Y_2
1.0
0.5
0.0
-0.5
-1.0
0.0
0.5
1.0
1.5
b
Y_2
Y_1

Fig. 4.

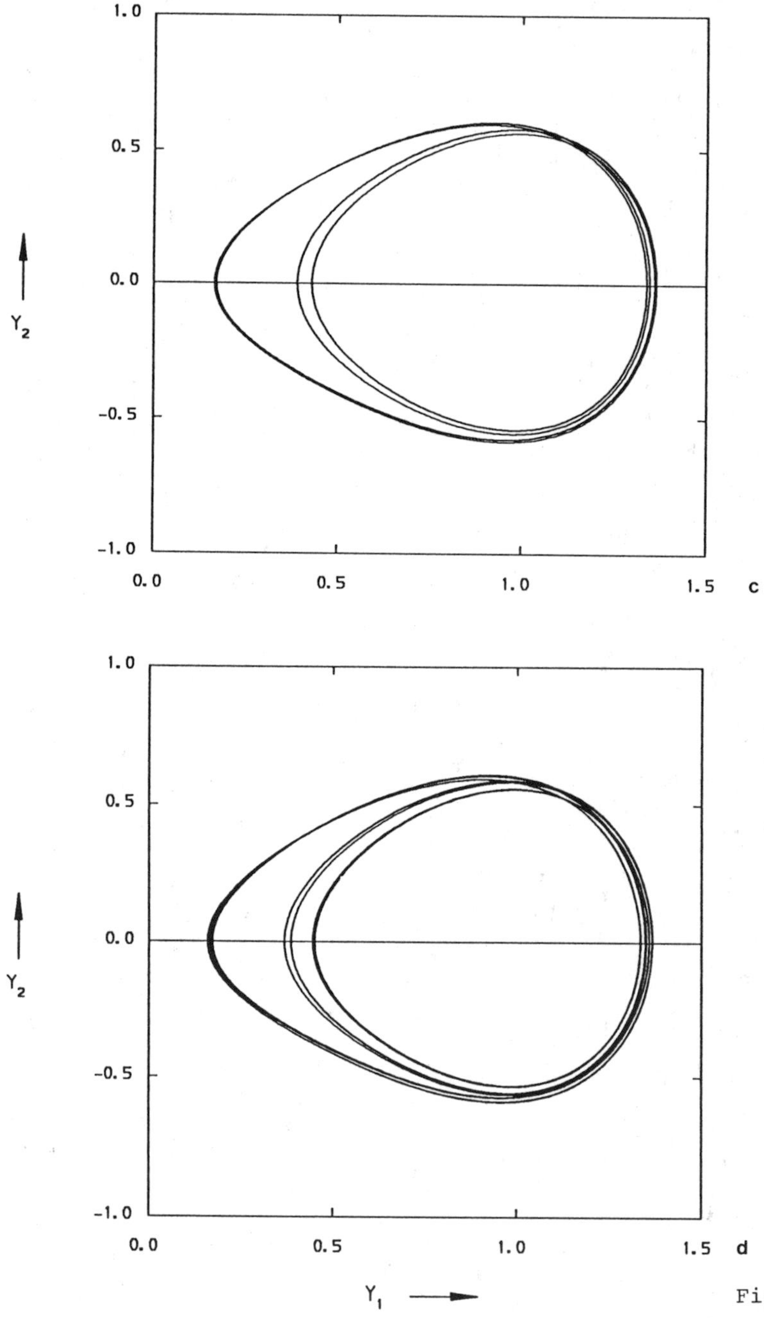

Fig. 4.

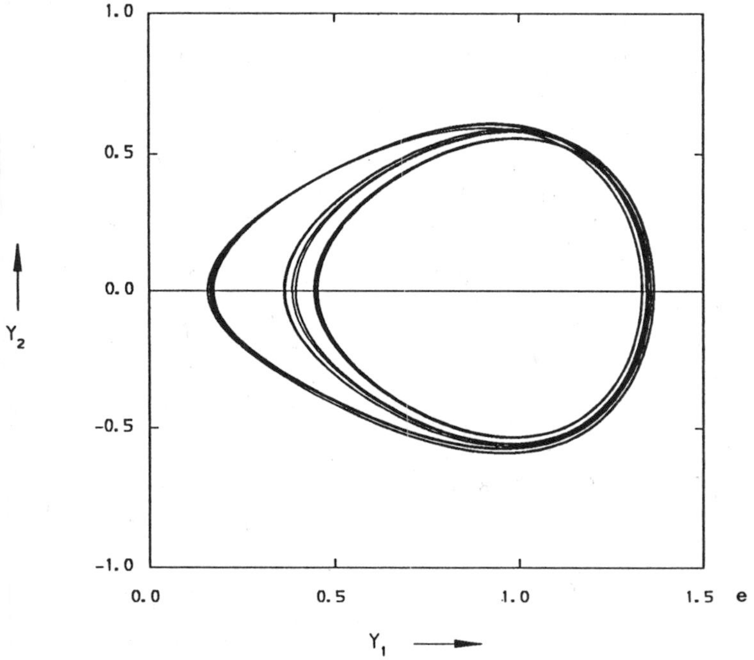

Fig. 4. Basic solution and successive subharmonics in the period doubling bifurcations leading to chaos: a) P = 0.07; b) P = 0.083; c) P = 0.0864; d) P = 0.08718; e) P = 0.08732.

For the other marked bifurcation values of P one of the multipliers crosses the unit circle at -1.

Four sequences of period-doubling bifurcations are shown and each new solution is stable over an interval smaller than the interval of the previous member of sequence. At the end chaos is reached (Fig. 5). This observation leads to another question: how compatible are presented results with the Feigenbaum universal constant [49]. Feigenbaum has analysed the noninvertible one-dimensional map $x_{n+1} = F_\lambda(x_n)$ dependent on a parameter λ. He has shown that the family F_λ has an infinite sequence of period-doubling bifurcations of

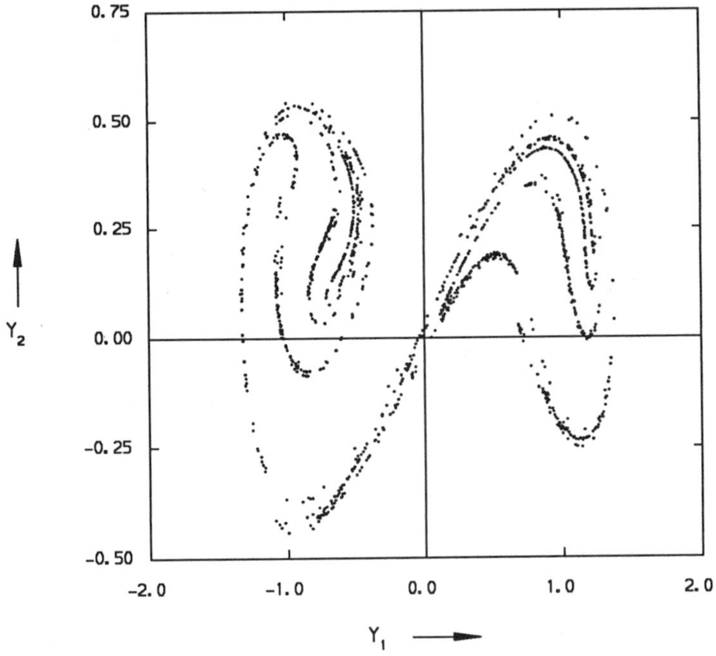

Fig. 5. The strange attractor as a limit of the period doubling sequence (P = 0.088).

periodic orbits as the parameter λ is increased. At a $\lambda \geqslant \delta$ chaos has appeared, where

$$\delta = \lim_{n\to\infty} \delta_n = \lim_{n\to\infty} \frac{\lambda_n - \lambda_{n-1}}{\lambda_{n+1} - \lambda_n} = 4.6692.$$

It should be pointed out, that because δ is independent of the exact nature of the function F_λ, this universal constant can be found for the physical systems governed by differential equations. This question was discussed by Tousi and Bajaj [50], where a system with two degrees of freedom has been studied. In their paper two sequences of period doubling bifurcations are presented. If one of the sequences seems to converge to δ, the other one seems to oscillate.

In our example, taking into account $\lambda = P$, we have obtained the following series δ_n: 8.076; 3.305; 18.154. Based on these results it seems that this scenario leading to chaos is not compatible with the Feigenbaum constant.

2.3.2. A particular oscillator with three equilibria

The investigated oscillator is governed by the equation

$$\ddot{y} + 0.25\dot{y} - y + y^3 = q + 0.4\cos t. \qquad (73)$$

This oscillator has unsymmetric elastic characteristics and a static load q. It is easy to check (see for instance [51], p. 23) that without periodic force the oscillator has three equilibrium positions, if only the inequality $q < 0.3$ is fulfilled. The equilibria are two sinks and a saddle. These types of oscillators were already investigated. Here attention is concentrated on equation (73) in order to present the new dynamical phenomena which accompanies the change of q. For q $(0.04657 \geqslant q \geqslant 0.0918)$ the 1/2 subharmonic solution has appeared (see Table 4). With further increase of q one of the characteristic multipliers crosses the unit circle at -1 and the new stable 1/4 subharmonic solution is born. With increase of q the absolute value of the first characteristic multiplier decreases, while the absolute value of the second characteristic multiplier increases and finally reaches the value of -1. Thus, for $q = 0.1103$ the bifurcation appears and a new 1/8 subharmonic stable solution is born.

Continuing with this procedure, the stable 1/16 subharmonic solution was found, which is stable for $(0.1103 \leqslant q \leqslant 0.1148)$. At the bifurcation point $q = 0.1148$ the 1/32 subharmonic solution is born.

The results presented above allow one to suppose that there

Table 4. Three successive subharmonic solutions for the oscillator (73).

q	y	$\dot{y}$	Multipliers 1	Multipliers 2	Period
0.04657 *	-0.34543	0.70910	0.04; 0.0	1. ; 0.0	2T = 4π
0.0466	-0.34599	0.70979	0.05; 0.0	0.95; 0.0	
0.0467	-0.34680	0.71073	0.05; 0.0	0.89; 0.0	
0.047	-0.34815	0.71220	0.05; 0.0	0.80; 0.0	
0.050	-0.35447	0.71792	0.11; 0.0	0.39; 0.0	
0.060	-0.36671	0.72585	-0.02; 0.21	-0.02;-0.21	
0.070	-0.37620	0.73035	-0.22; 0.0	-0.20; 0.0	
0.080	-0.38460	0.73351	-0.66; 0.0	-0.07; 0.0	
0.090	-0.39234	0.73584	-0.95; 0.0	-0.05; 0.0	
0.0918 ▪	-0.39368	0.73620	-1. ; 0.0	-0.04; 0.0	
0.0930	-0.40051	0.74020	0.89; 0.0	0.0 ; 0.0	4T = 8π
0.0940	-0.40348	0.74169	0.78; 0.0	0.0 ; 0.0	
0.0950	-0.40587	0.74279	0.68; 0.0	0.0 ; 0.0	
0.0960	-0.410795	0.74369	0.57; 0.0	0.0 ; 0.0	
0.0970	-0.38236	0.72634	0.46; 0.0	0.0 ; 0.0	
0.0980	-0.38151	0.72533	0.36; 0.0	0.0 ; 0.0	
0.0990	-0.41326	0.74575	0.24; 0.0	0.01; 0.0	
0.1000 *	-0.41483	0.74630	0.13; 0.0	0.01; 0.0	
0.103	-0.37845	0.72105	0.02; 0.0	-0.20; 0.0	
0.105	-0.37757	0.71956	0.01; 0.0	-0.42; 0.0	
0.108	-0.37644	0.71747	0.01; 0.0	-0.74; 0.0	
0.110	-0.37581	0.71626	0.01; 0.0	-0.96; 0.0	
0.1103 ▪	-0.37572	0.71597	0.01; 0.0	-1.0 ; 0.0	
0.1124	-0.38167	0.72077	0.05; 0.24	0.05;-0.24	8T = 16π
0.113	-0.38268	0.72144	-0.35; 0.0	0.06; 0.0	
0.114 *	-0.38418	0.72239	-0.49; 0.0	-0.20; 0.0	
0.1148	-0.38527	0.72304	-1.0 ; 0.0	-0.11; 0.0	

exists an infinite sequence of period-doubling bifurcations which finally lead to chaos. For the marked values of q in Table 4, phase portraits of three successive subharmonic solutions are presented in Fig. 6. For q = 0.12 the strange chaotic attractor is shown in Fig. 7.

It should be pointed out, that when each of the demonstrated subharmonic solution has lost its stability, another subharmonic solution with twice the previous period appeared. Secondly, the lives of the higher subharmonic solutions became shorter when q was increased. On the other hand, unstable subharmonic solutions exist for all of the orders in a very short range of the parameter q. Finally, we can suppose that in the interval of q, where chaotic motion is observed, no stable solutions exist, but infinitely many unstable successive subharmonics.

Consider again the behaviour of the 1/2 subharmonic solution when q decreases starting with the value 0.0918. The second multiplier (see Table 4) increases and goes first through the real negative, than through imaginary and real positive values. Finally, for q = 0.04657 it reaches +1 and further progress can not be made for $q > q^*$, with the Newton method, since Newton's procedure fails to converge. Poincaré map for q = 0.0465 has been demonstrated in Fig. 8 which explains the sudden appearence of the strange attractor.

2.3.3. Oscillator with a particular exciting force

The third considered nonlinear sinusoidally driven oscillator is governed by the equation

$$\ddot{y} - (\beta - \delta y^2)\dot{y} + \alpha y + \mu y^3 = q + \eta^2 \cos\eta t. \qquad (74)$$

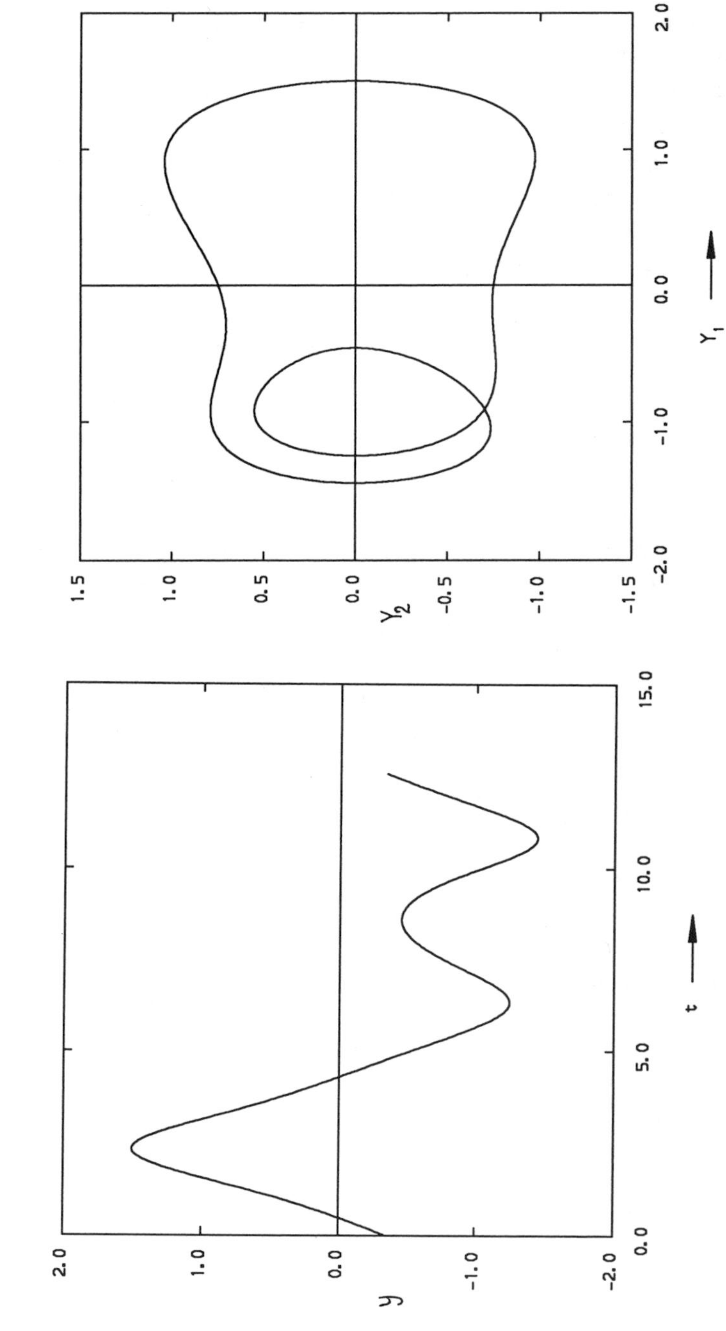

2.0
1.0
0.0
-1.0
-2.0
Y1
1.5
1.0
0.5
0.0
-0.5
-1.0
-1.5
Y2
0.0
5.0
10.0
15.0
t
2.0
1.0
0.0
-1.0
-2.0
y

Fig. 6. a

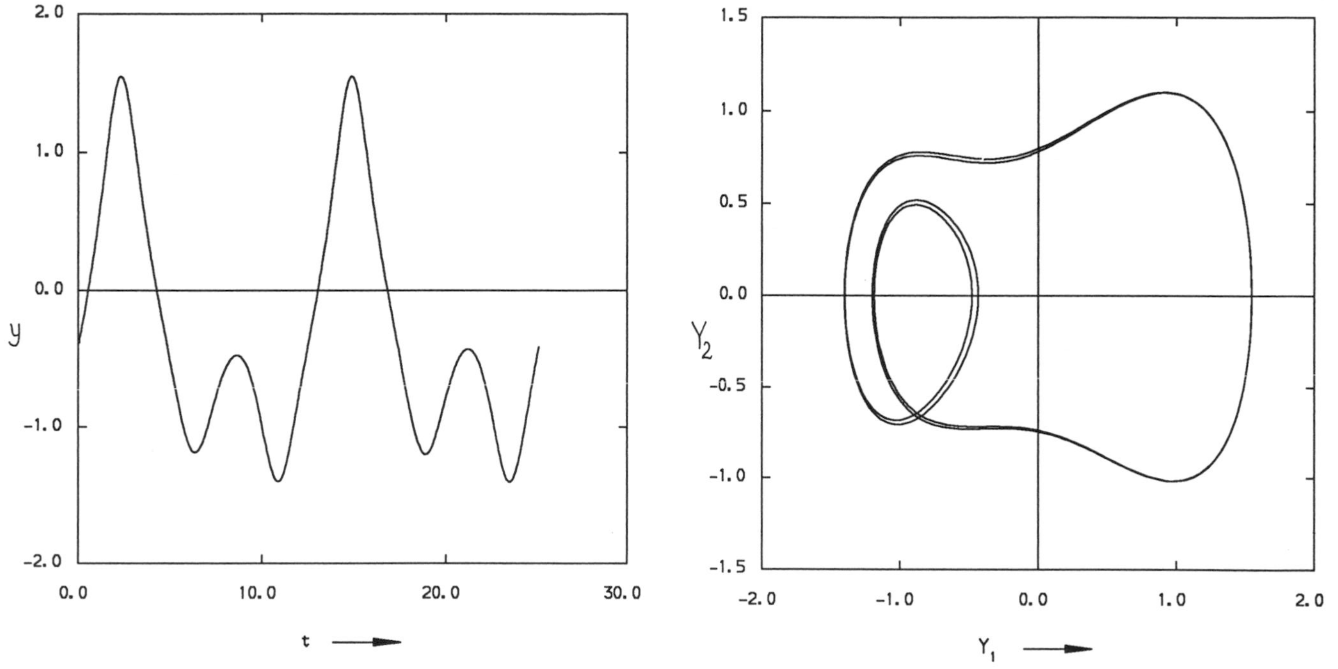
2.0
1.0
0.0
-1.0
-2.0
y
0.0
10.0
20.0
30.0
t
1.5
1.0
0.5
0.0
-0.5
-1.0
-1.5
Y2
-2.0
-1.0
0.0
1.0
2.0
Y1

b

Fig. 6.

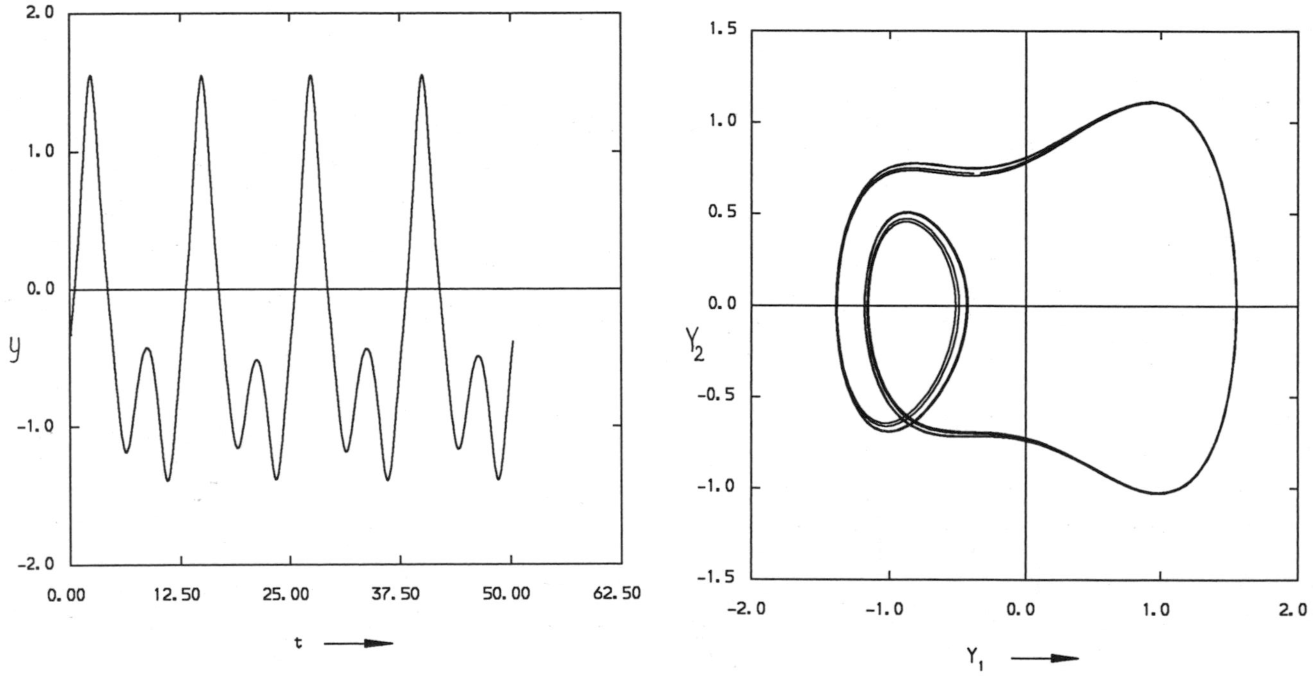

c

Fig. 6. Period doubling scenario: a) q = 0.04657; b) q = 0.1; c) q = 0.114.

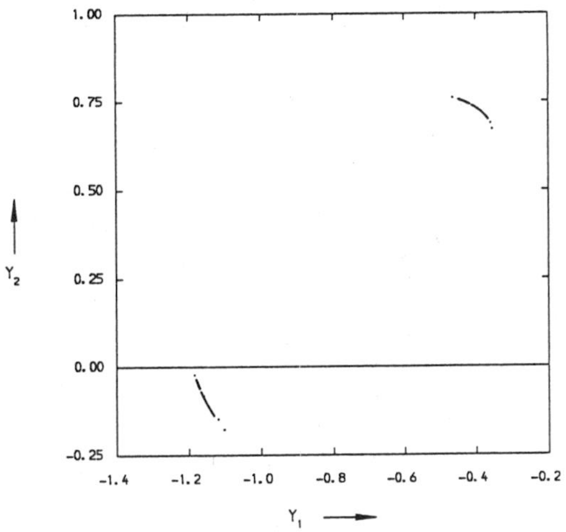

Fig. 7. Poincaré map of a chaotic attractor for q = 0.12.

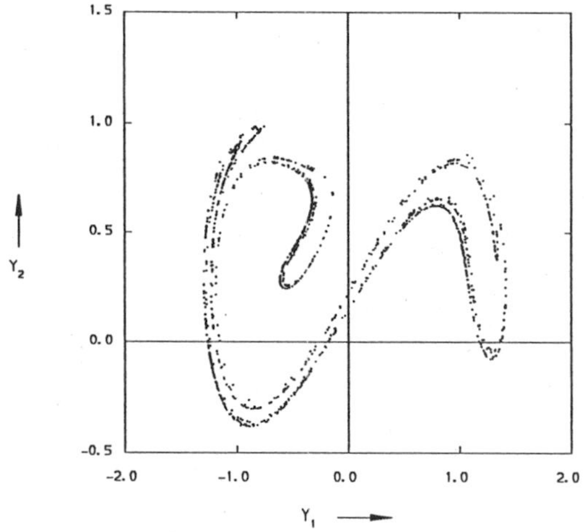

Fig. 8. The strange attractor after a multiplier crosses the unit circle at +1 (q = 0.0465).

This oscillator has an exciting force whose amplitude is proportional to the second power of the exciting frequency. The motion of the mechanical system with one degree of freedom, where the rotation of an unbalanced disk or wheel becomes the source for the exciting force, can be reduced to the equation (74). This oscillator has one equilibrium position and its chaotic behaviour supported by approximate analytical methods will be investigated in Chapter 3.

Numerical results, similar to those in the previous two sections, are presented in Table 5.

Table 5. Periodic solution of the oscillator (74).

β	y	$\dot{y}$	Multipliers		Period
			1	2	
-0.1	1.4976	0.075746	-0.78; 0.48	-0.78;-0.48	
-0.08	1.4979	0.066015	-0.79; 0.49	0.79;-0.49	
-0.05	1.4983	0.051408	-0.80; 0.49	-0.80;-0.49	
-0.03	1.4984	0.041665	-0.81; 0.50	-0.81;-0.50	
0.0	1.4986	0.027044	-0.82; 0.50	-0.82;-0.50	$T = 2\pi$
0.05	1.4988	0.0026691	-0.83; 0.51	-0.83;-0.51	
0.08	1.4988	-0.011956	-0.84; 0.52	-0.84;-0.52	
0.09 *	1.4988	-0.016831	-0.85; 0.52	-0.85;-0.52	
0.098	1.4987	-0.020730	-0.85; 0.52	-0.85;-0.52	
0.1	1.4987	-0.021705	-0.85; 0.53	-0.85; 0.53	

Following parameters are fixed: $\delta = 0.11$, $\alpha = 0.1$, $\mu = 69.17$, $q = 1.0$, $\eta = 7.8$. We start with the value $\beta = -0.1$, where a periodic orbit with the period 2π was found. Two characteristic multipliers are complex conjugates. With increase of β these two complex conjugate eigenvalues move outside of the unit circle. The phase portrait of the system close to the loss of stability is presented in Fig. 9.

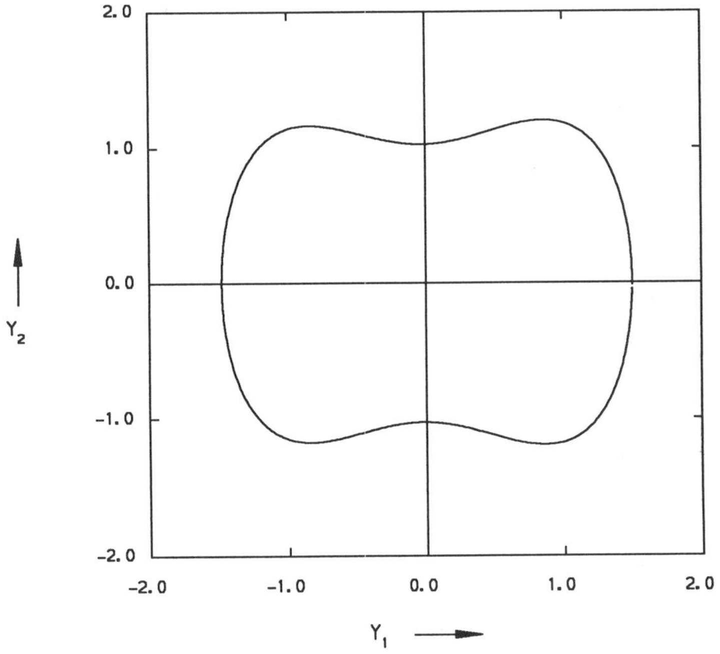

Fig. 9. Periodic orbit of the oscillator (74) near the loss of stability ($\beta = 0.09$).

For $\beta = 0.1$ two multipliers lie on the radius with the length 1.0017. It means that the previous stable periodic orbit becomes unstable.

The problem of loss of stability of a periodic orbit is of degeneracy 1 (see Arnold [52]). However, if a pair of multipliers crosses the unit circle, generally two essential parameters are needed to analyse this case. This situation has also been considered by Aronson et al. [53], where their system was embeded in a two-parameter family. They have shown, that for certain parameter values a Hopf bifurcation to an invariant circle appears, which is smooth for the parameter values in a neighbourhood of the bifurcation point. For the parameter values far from the bifurcation point the investigated invariant set fails to be even topologically a circle and chaos appears.

Secondly, following Arnold, in the considered case, topologically versal deformations are not known and may not exist.

The numerical example presented here shows that for $\beta = 0.1$ the strange attractor has appeared (Fig. 10). The two above-cited works indicated this possibility.

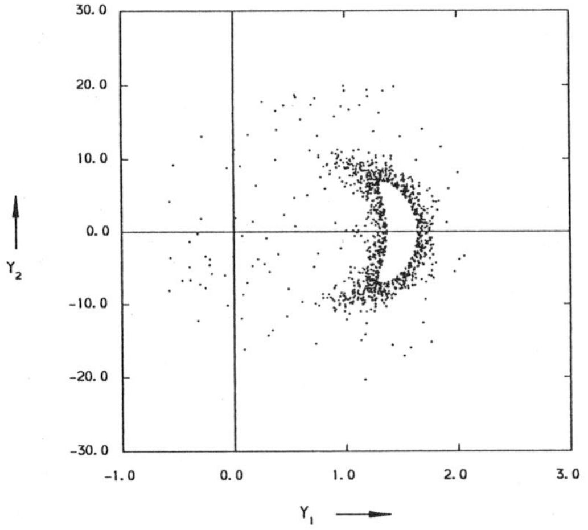

Fig. 10. The strange attractor obtained after a pair of multipliers have crossed the unit circle ($\beta = 0.1$).

2.3.4. Concluding remarks

Three particular sinusoidally driven oscillators serve as examples to present new dynamical phenomena when the bifurcation of periodic orbits occurs. Following Arnold, different types of bifurcation are distinguished by the way one eigenvalue (or a pair of eigenvalues) crosses the unit circle of the complex plane. Birth or annihilation of periodic orbits appear when one multiplier crosses the unit circle at +1. The new subharmonic solution is born when one multiplier crosses the unit circle at -1. Quasiperiodic or another resonance solutions appear, when a pair of multipliers cross the unit circle simultaneously. The presented examples show, however, that for the two above-mentioned cases (passage of a pair of multipliers through the unit circle and one eigenvalue through +1) the previous attracting periodic orbit vanishes and chaotic orbits appear instead.

2.4. Bifurcation of the Oscillations of Vocal Cords

The investigated nondimensional nonlinear differential equations modelling the vibrations of the human vocal cords were established by Cronjaeger [54]. It can be cast in the following form:

$$\ddot{x} + d\dot{x} + (k_x + k_c((x - X_0)^2 + y^2))(x - X_0) +$$

$$- k_{xy}y - k_s x^{-4}(1 - d_s\dot{x}) = Ep, \qquad (75)$$

$$\ddot{y} + d\dot{y} + (k_y + k_c((x - X_0)^2 + y^2))\, y - k_{xy}\,(x - X_0) = Ep,$$

$$\dot{p} = Q - \begin{cases} (x - 1)p^{1/2} & \text{for } x > 1, \\ 0 & \text{for } x \leqslant 1, \end{cases}$$

where d is damping of the vocal cords, $k_x(k_y)$ is a horizontal (vertical) stiffness of the vocal cords, k_{xy} is a stiffness of the couplings between two directions of motion, k_c is cubic type stiffness, k_s is hyperbolic type stiffness, d_s — damping, X_0 — unloaded equilibrium position (Q = 0), E — average pressure, Q — air flow.

Coordinates x, y are horizontal and vertical displacements of the vocal cords, while p is the air pressure. Among the ten parameters the following are fixed: $k_x = 1$, $k_y = 0.3$, $k_c = 0.001$, $k_s = 0.001$, $d_s = 0.5$, $k_{xy} = 0.3$, $X_0 = 0.4$, $E = 0.4$, $Q = 7$. For convenience we abbreviate the equations of motion by

$$\underline{h}(\underline{z}', \underline{z}, \eta) = \underline{0}, \tag{76}$$

where: $\underline{z} = (x, \dot{x}, y, \dot{y}, p)$ and η is vector of parameters (further details see in [55]). For the constant solution $\underline{z} = \underline{z}_c$ we obtain

$$\underline{h}(\underline{0}, \underline{z}_c, \eta_0) = 0. \tag{77}$$

Equation (77) was solved numerically by Newton's method. From (76) one can obtain the linear first order variation equation

$$\Delta\underline{z}' = \underline{H}\,\Delta\underline{z}, \tag{78}$$

where H is a constant matrix and Δz is a vector of small perturbations of the constant solution. The eigenvalue problem of (78) yields the five eigenvalues and corresponding eigen-

During the calculations, the frequency ω and relative time $\tau = \omega t$ is introduced. Frequency ω enters the equations as a parameter to be kept fixed when we look at a special solution. Periodic solutions z(t) with the period T correspond to periodic solutions $z(\tau, \omega)$ with the period 2π. Because the system (75) is autonomous a phase condition $\dot{x} = 0$ is prescribed in order to fix the unknown frequency ω. If the solution is unfavorably stated (not uniquely solvable for $\underline{z}$ (ω)), an exchange of the fastest varying z_1 variable removes the numerical difficulties (z_1 is chosen, ω calculated). During numerical integration because of the nonlinear term x^{-4}, the standard methods based on the Runge-Kutta algorithm are not sufficiently accurate to solve the considered stiff equations system. We have used a variable-order, variable-step Gear method with a very small step for $x \cong 0$ and calculations were made with double precision.

Figure 11 shows the branching diagram for damping $0.01 \leqslant d \leqslant 0.35$. The previous stable steady-state solution becomes unstable for d = 0.35 and the periodic limit cycle emanates from Hopf bifurcation point H1. At this point a new periodic solution has a frequency $\omega = 1.169$ which is equal to the imaginary eigenvalue obtained from (78). Then the period of this solution is normalized to 2π and the branch H1PD1 is obtained. As can be seen from the corresponding diagram with multipliers (Fig. 12) for d = 0.295 the real eigenvalue crosses the unit circle at -1 and the second subharmonic solution has appeared with $\omega = 1.197$. The previous periodic solution becomes unstable. However, with the further decrease of damping the real eigenvalue turns to the right and for d = 0.19 (ω = 1.332) crosses the unit circle at -1 again. After this, all eigenvalues are inside the unit circle and the considered solution marked as 1 is stable. Points PD2 (d = 0.19, ω = 1.332) and PD3 (d = 0.1, ω = 1.402) belong to the envelope of self-excited oscillations which correspond to the second unstable subharmonic solution (coded as 4). After crossing the point PD3 the considered 2π periodic solution becomes unstable again. From the point PD1 branches another subharmonic solution. This solution has period 4π, but during the calcu-

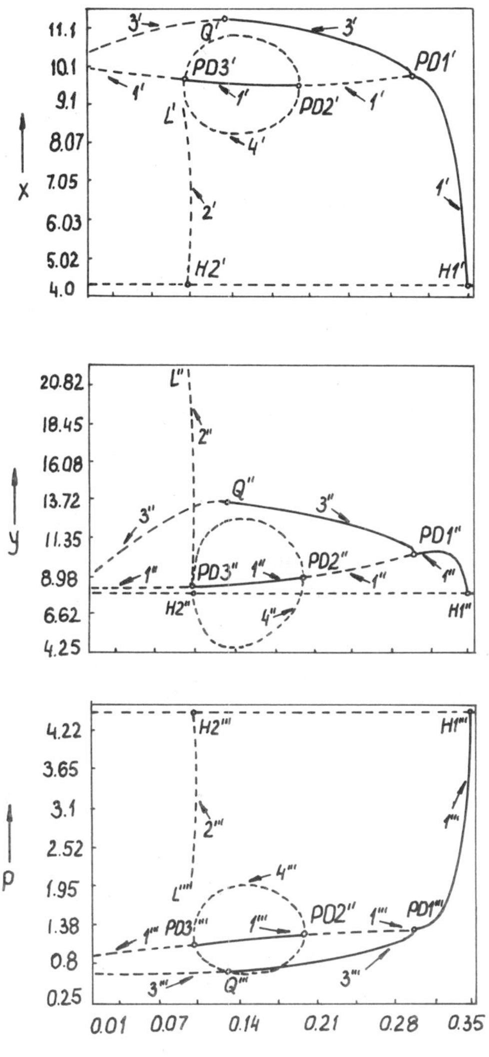

Fig. 11. Bifurcation portrait of oscillations of the human vocal cords. At the bifurcation points H1, H2 and PD1 only one branch of the new emanating solutions are marked in order to present the phase shift between demonstrated variables $y_1 = x$, $y_3 = y$, $y_5 = p$. The solid line indicates stable solutions.

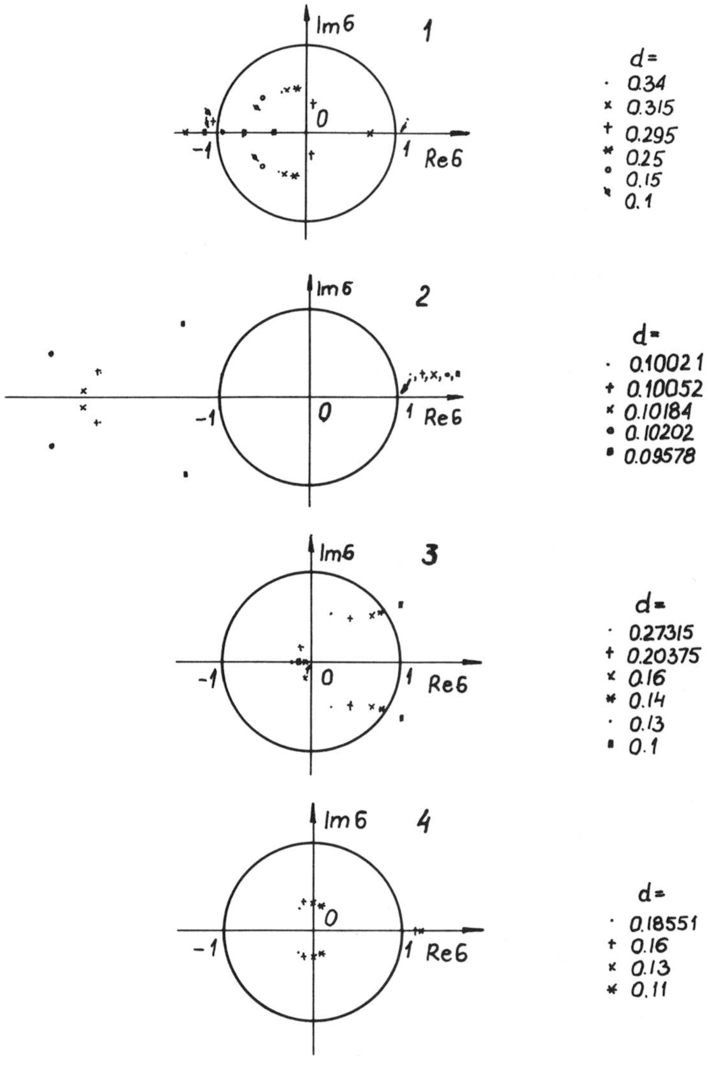

Fig. 12. The travel of the characteristic multipliers by varying damping d. Each of the four examples correspond to the marked branches in Fig. 9 (only three multipliers are presented).

lations it is normalized to 2π with $\omega = 0.608$ at the point PD1. With decrease of d, frequency ω increases and for d = 0.01 reaches value 0.695.

With the decrease of d, two complex conjugate eigenvalues approach the unit circle and for d = 0.13 cross it. The subharmonic solution marked as 3 becomes unstable and the situation does not change in the considered interval of the parameter d. Another solution branches from the point Q (d=0.13). The question arises if this solution is periodic or quasiperiodic. The two conjugate multipliers σ_* lie on the unit circle, when $\sigma_* = e^{2\pi\omega_* i}$, and $\omega_* = \frac{1}{2\pi}\arg(\sigma_*)$. In this case the eigenvalues lying on the unit circle are (0.80, $\pm$0.60) and $\omega_* \neq l/k$, where l, k are relatively prime integers. It means that at this point a new quasiperiodic solution is born.

From the second Hopf bifurcation point H2 branches the periodic solution 2 which is unstable. For some narrow interval for the fixed d value there exist two unstable periodic solutions. From the last numerically obtained point (L) belonging to the curve 2, further progress of the numerical calculation is impossible. The solution has left the real plane. Calculated stable and unstable limit cycles are shown (Fig. 13) for such values of d which belong to all the separated branches.

To conclude, the calculation method to obtain bifurcation diagrams of the 5th order system of differential nonlinear equations has been presented. This system of equations describes the self-excited vibrations of the vocal cords. For the considered interval of damping it has been shown that this system possesses two Hopf bifurcation points, three period doubling bifurcation points and one Q point. After Q is crossed, a quasiperiodic torus is born. There is an interval of damping for which two stable 2π periodic and 4π periodic solutions exist. It is possible to jump from one to another solution. Harmonic and subharmonic unstable solutions also exist for this considered interval of damping.

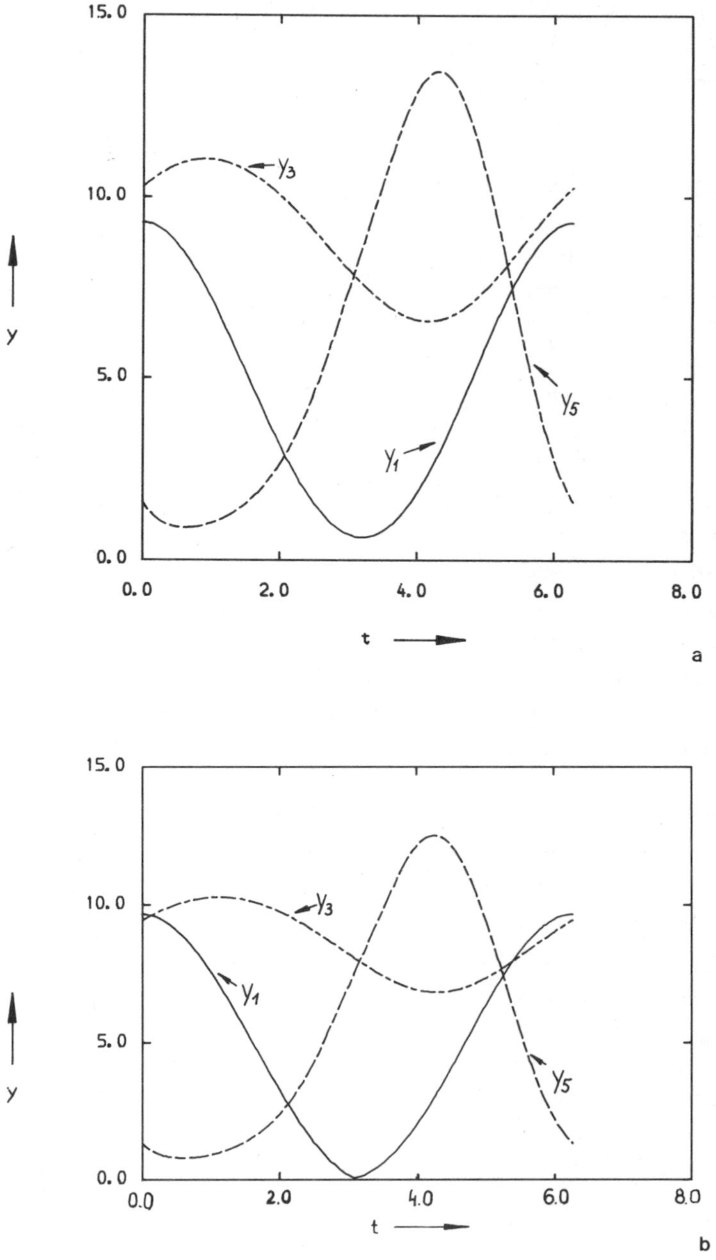
15.0
10.0
5.0
0.0
y
y3
y5
y1
0.0
2.0
4.0
6.0
8.0
t
a
15.0
10.0
5.0
0.0
y
y3
y1
y5
0.0
2.0
4.0
6.0
8.0
t
b

Fig. 13.

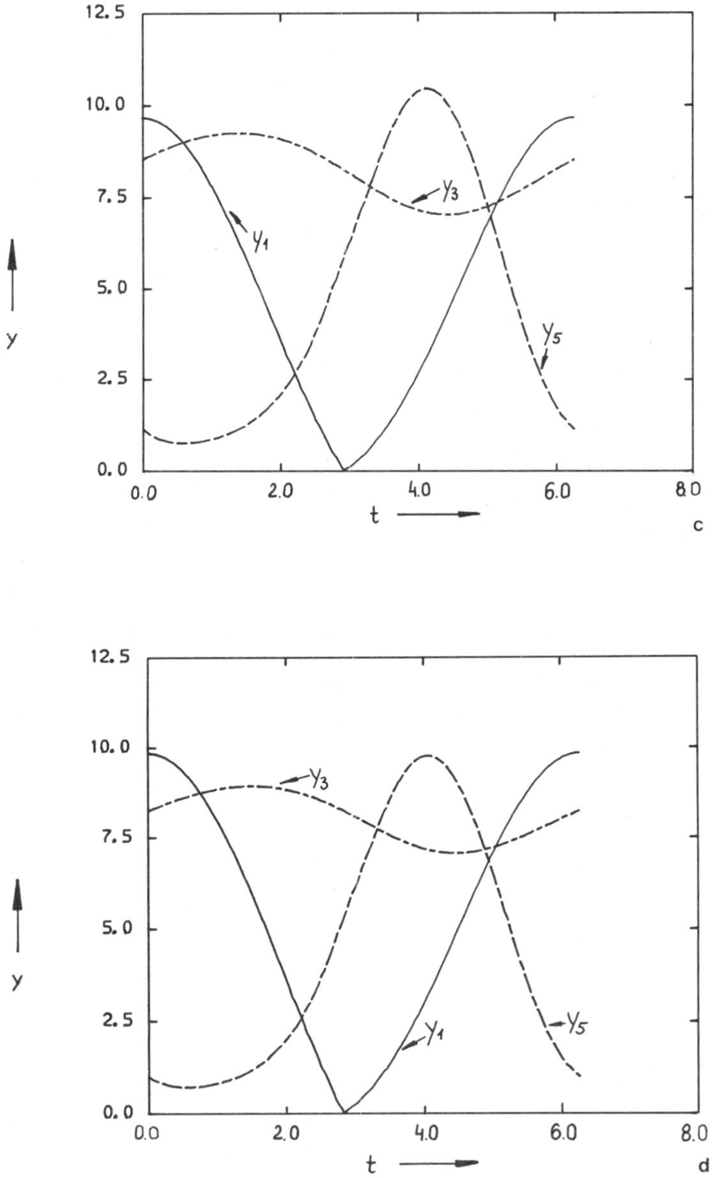
12.5
10.0
7.5
5.0
2.5
0.0
y
0.0
2.0
4.0
6.0
8.0
t
y_1
y_3
y_5
c
12.5
10.0
7.5
5.0
2.5
0.0
y
0.0
2.0
4.0
6.0
8.0
t
y_1
y_3
y_5
d

Fig. 13.

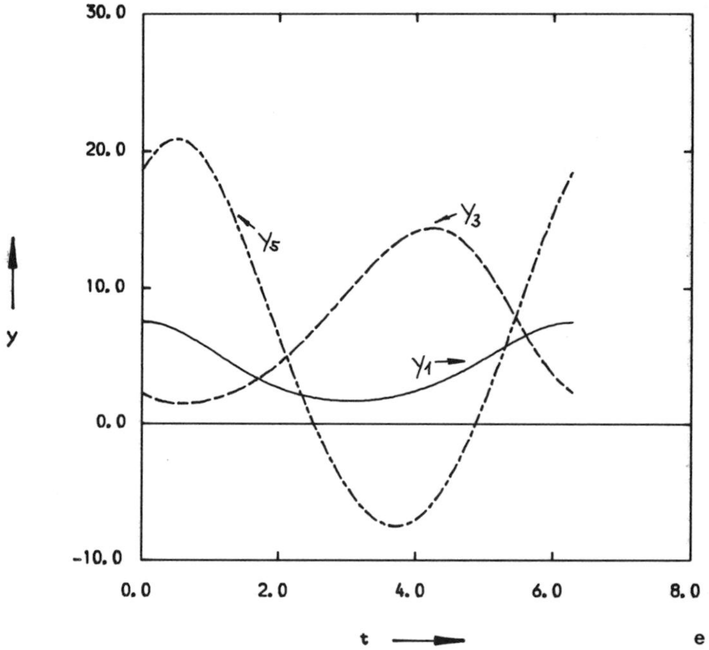
30.0
20.0
10.0
0.0
-10.0
y
0.0
2.0
4.0
6.0
8.0
t
y_5
y_3
y_1
e

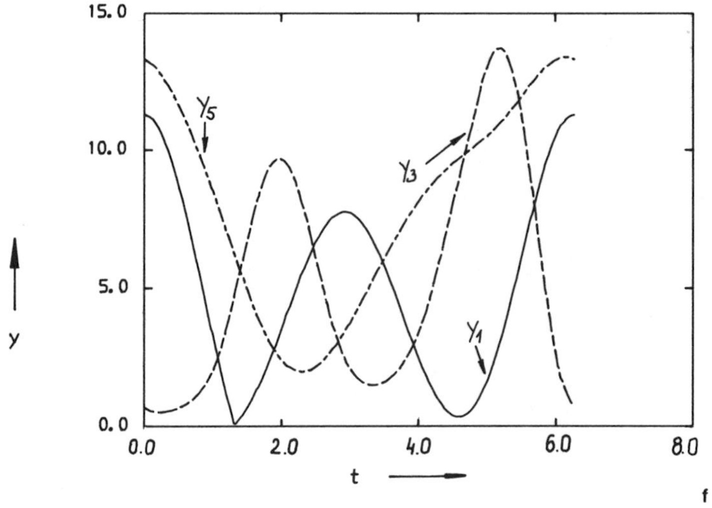
15.0
10.0
5.0
0.0
y
0.0
2.0
4.0
6.0
8.0
t
y_5
y_3
y_1
f

Fig. 13.

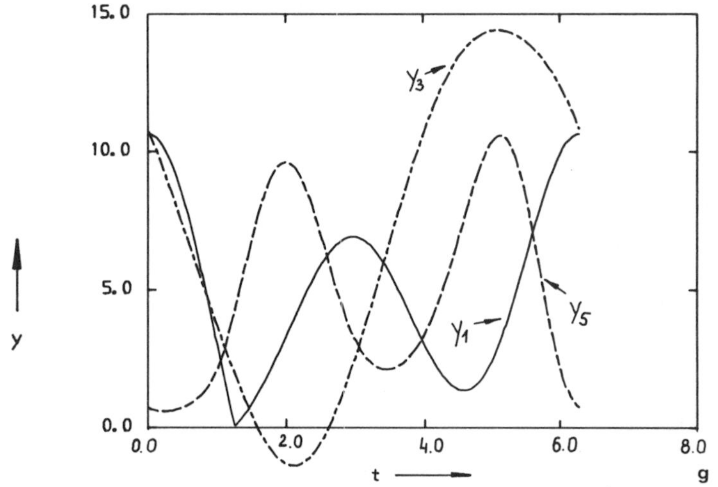

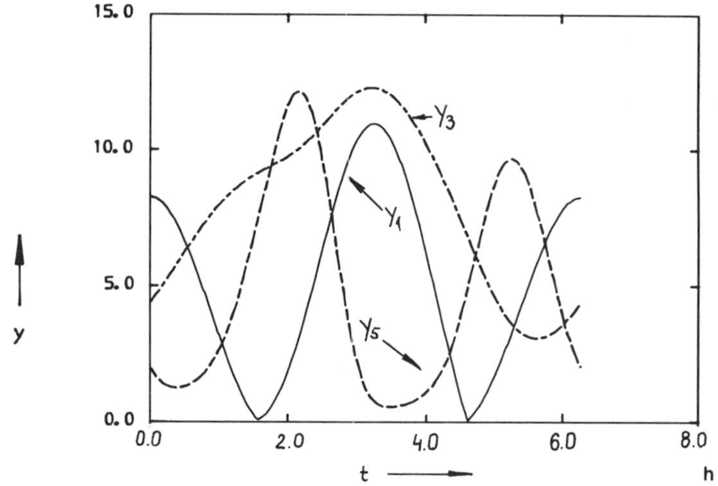

Fig. 13. Calculation examples of the stable and unstable limit cycles. Time histories of oscillations of the vocal cord correspond to the branches from Fig. 11 as follows:(a,b,c,d) to 1;(e) to 2;(f,g) to 3 and (h) to 4. The calculations were made for the following values of damping d: (a) d=0.32;(b) d=0.25; (c) d=0.13; (d) d=0.06; (e) d=0.1;(f) d=0.16;(g) d=0.04; (h) d=0.13.

Chapter 3

3. CHAOS AFTER BIFURCATION OF PERIODIC AND QUASIPERIODIC ORBITS

3.1. Introduction

As has been shown in Chapter 2, the bifurcation of periodic orbits appears when the multipliers have values of +1, -1 or a pair of complex conjugate eigenvalues simultaneously cross the unit circle of the complex plane. The last possibility is known as a Hopf bifurcation for periodic orbits [56], because thanks to the use of the cross-section of the investigated orbit and return map one can consider a fixed point instead of a full periodic orbit. In this sense a theory of simple bifurcations of periodic orbits is analogous to the theory of equilibria. Analytical approximate approach allows one to obtain averaged Hopf bifurcation equations for periodic (or quasiperiodic) orbits. Examples given below show also that by using this analytical technique chaos can be found. Results presented here are based on references [57-62].

3.2. Oscillator with a Static Load and Particular Exciting Force

3.2.1. Bifurcation of periodic orbit with one frequency

An oscillator governed by the dimensionless equation

$$\ddot{y} - (\beta - \delta y^2)\dot{y} + \alpha y + \mu y^3 = q + \eta^2 \cos\eta t \tag{81}$$

is considered.
Assuming that the stationary solution has the form

$$y = Y + A\cos\eta t + B\sin\eta t, \tag{82}$$

we obtain from (81)

$$Y\,(\alpha + \mu(Y^2 + \tfrac{3}{2}P^2)) - q = 0,$$

$$B\,(\alpha - \eta^2) - A\eta(-\beta + \delta(Y^2 + \tfrac{1}{4}P^2)) + 3\mu B(Y^2 + \tfrac{1}{4}P^2) = 0,$$

$$A\,(\alpha - \eta^2) + B\eta(-\beta + \delta(Y^2 + \tfrac{1}{4}P^2)) + 3\mu A(Y^2 + \tfrac{1}{4}P^2)$$

$$-\eta^2 = 0, \tag{83}$$

where $P = (A^2 + B^2)^{1/2}$ is the amplitude of vibration. The perturbated solution of (82) is

$$y_p = Y + \Delta Y + (A + \Delta A)\cos\eta t + (B + \Delta B)\sin\eta t. \tag{84}$$

Assuming that perturbations $\Delta(\cdot)$ and damping coefficients β and δ are small, we obtain

$$-2\eta(\dot{\Delta A}) + T(\Delta A) + U(\Delta B) = 0,$$

$$2\eta(\dot{\Delta B}) + V(\Delta A) + W(\Delta B) = 0, \tag{85}$$

where:

$$T = \beta\eta - \eta\delta Y^2 - \tfrac{3}{4}\delta\eta A^2 - \tfrac{1}{4}\delta\eta B^2 + \tfrac{3}{2}\mu BA,$$

$$U = \alpha - \eta^2 + 3\mu Y^2 + \tfrac{3}{4}\mu A^2 + \tfrac{9}{4}\mu B^2 - \tfrac{1}{2}\delta\eta AB,$$

$$V = \alpha - \eta^2 + 3\mu Y^2 + \tfrac{9}{4}\mu A^2 + \tfrac{3}{2}\mu B^2 + \tfrac{1}{2}\delta\eta AB, \tag{86}$$

$$W = -\beta\eta + \delta\eta Y^2 + \tfrac{1}{4}\delta\eta A^2 + \tfrac{3}{4}\delta\eta B^2 + \tfrac{3}{2}\mu AB.$$

At the bifurcation point we have

$$W - T = 0, \tag{87}$$

and

$$UV - TW > 0. \tag{88}$$

Consequently, we obtain the set of equations (83), (87) and the inequality (88). In order to obtain the bifurcation curves $\mu(\eta)$ we solve those equations for arbitrarily chosen parameters α, β, δ, q. Suppose, that we have found the Hopf bifurcation curve $\mu(\eta)$. For the parameters lying on the one side of this curve the real parts of the eigenvalues of (85) are negative (positive) whereas for the other side they are positive (negative). For each point belonging to this curve we have a corresponding stationary state with Y_H and amplitudes A_H, B_H. After crossing this curve from negative real parts to positive real parts of the parameter plane with non-zero velocity the Hopf bifurcation appears. Our considerations are valid for the averaged system of the equations after substituting (82) and (84) in to (81), and additionally, we investigate only the local bifurcation of Hopf type (see also [63]) This bifurcation in the averaged system of equations is related to the bifurcation of the solution (82) in equation (81). We finally obtain the conditions necessary for a bifurcation of the stationary state with one frequency.

Solutions of (83) and (87) are found using Newton's method. Sample curves obtained in this way are shown in Fig. 14.

The considered system of algebraic nonlinear equations can possess one, two or three equilibrium paths. It depends on the parameter values. In Figure 14 we show that for $\alpha = 0.1$ and $\alpha = 10.0$ we have, in each case, two bifurcation curves. For parameters near the critical values, we integrate equation (81) numerically by the Runge-Kutta method with the step size $h = 2\pi/(50\eta)$ and with the initial conditions $y(0) = 0.0$, $\dot{y}(0) = 1.0$. The results are presented in the form of

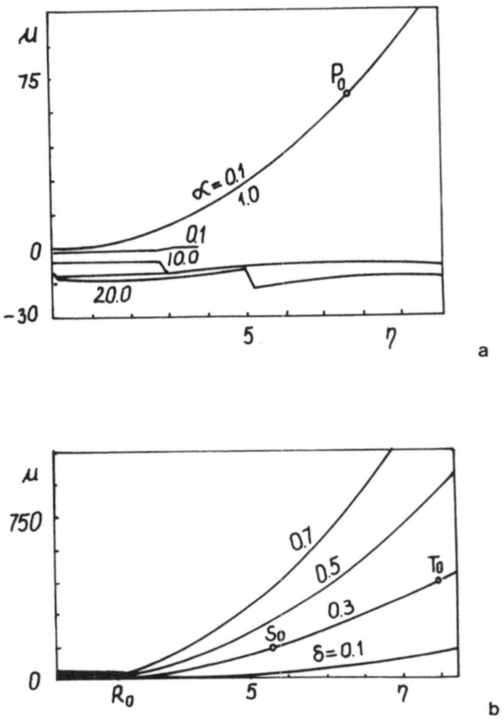

Fig. 14. Bifurcation curves for the one - frequency solution
a) $\beta = 0.1$, $\delta = 0.1$, $q = 1.0$;
b) $\beta = 0.1$, $\alpha = 0.1$, $q = 3.0$.

Poincaré maps and frequency spectra. The Poincaré maps consist of 2000 points. These maps are calculated starting with time $t = 100T$ ($T = 2\pi/\eta$), for which the trajectories finally reach the attractor. The Fourier spectra are obtained using Fast Fourier Transform and are presented in three-dimensional form. A decimal scale has been adopted for amplitudes, corresponding frequencies and time. The solutions have been analysed starting from $t = 100T$ and then we have obtained the Fourier components of the motion repeating the analysis of each interval of $300T$. This method of calculation of the Fourier spectra allows us to observe the development of motion and to trace the evolution of the Fourier components with time.

Consider the behaviour of the system near the bifurcation point P_0 in the Fig. 14. Figure 15 demonstrates sudden changes in nonchaotic and chaotic dynamics as parameter β is varied. For $\beta = 0.08$ we can observe very long, transient oscillations which finally reach the nonchaotic attractor, indicated by the stable fixed point F on the Poincaré map. After a long transient state, a periodic motion with the frequency of the exciting force remains (Fig. 15a). When β is slightly increased (Fig. 15b; $\beta = 0.085$) transient chaos occurs. The development of the power spectrum testifies to this. The chaotic motion of the system lasts until t = 800T. In the time trajectory the two broad-band regions of frequency near of

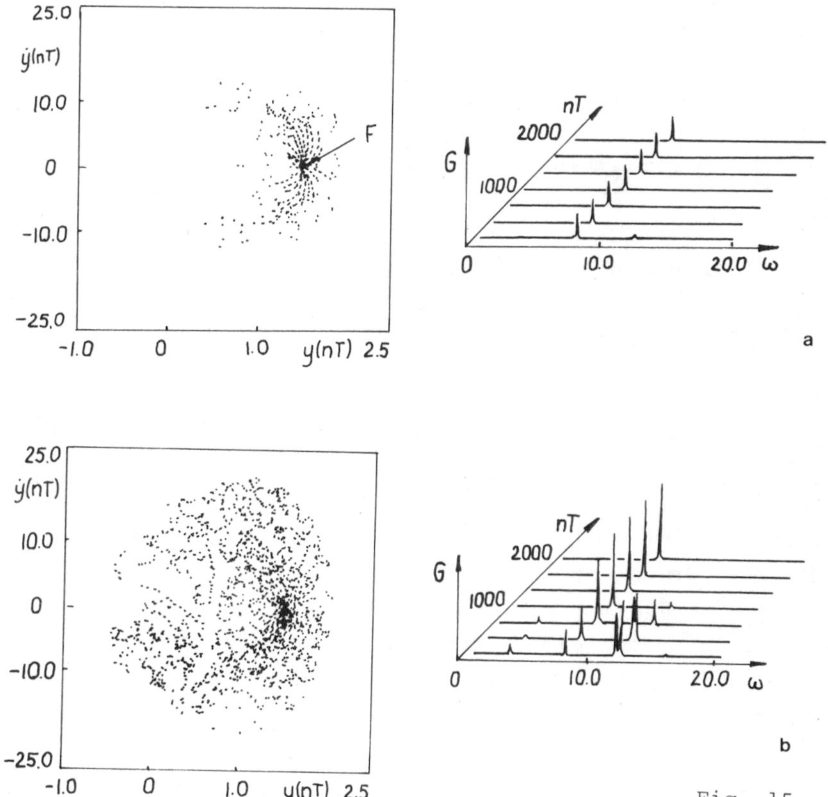

Fig. 15.

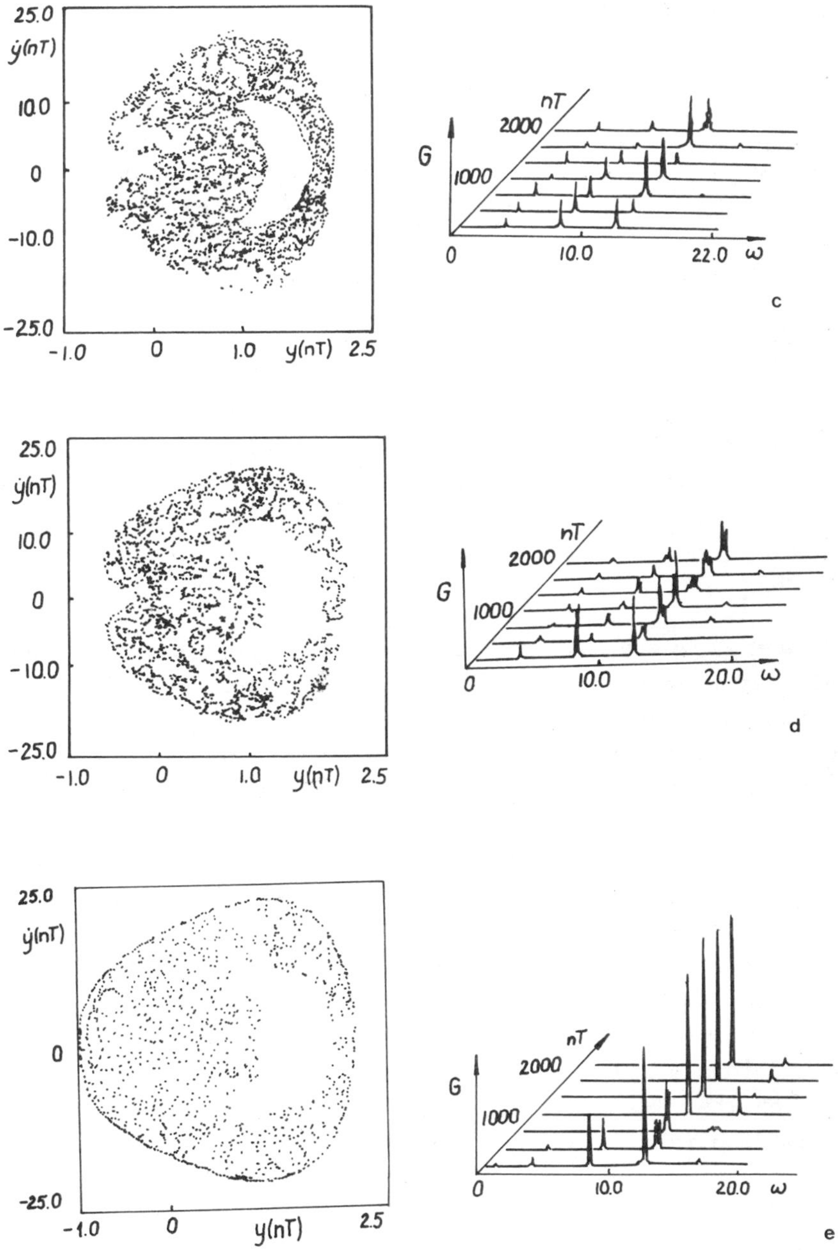
25.0
ẏ(nT)
10.0
0
-10.0
-25.0
-1.0
0
1.0
y(nT)
2.5
nT
2000
1000
G
0
10.0
22.0
ω
c
25.0
ẏ(nT)
10.0
0
-10.0
-25.0
-1.0
0
1.0
y(nT)
2.5
nT
2000
1000
G
0
10.0
20.0
ω
d
25.0
ẏ(nT)
0
-25.0
-1.0
0
y(nT)
2.5
nT
2000
1000
G
0
10.0
20.0
ω
e

Fig. 15.

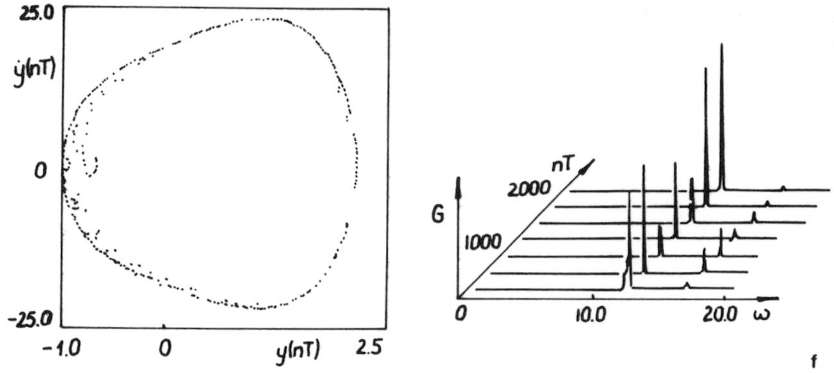

Fig. 15. Evolution of attractors with change of β and corresponding Fourier spectra: $\alpha = 0.1$; $\delta = 0.1$; $\mu = 69.17$ $\eta = 7.8$; $q = 1.0$: a) $\beta = 0.08$; b) $\beta = 0.085$; c) $\beta = 0.09$; d) $\beta = 0.095$; e) $\beta = 0.098$; f) $\beta = 0.1$.

$1/2\,\eta$ and $2/3\,\eta$ are evident. The transient chaotic motion then gradually changes into a regular two-frequency motion with frequencies η and 2η. A clearly dominating component is the frequency of the exciting force. With further increase of β, the transitional phenomena become longer and suddenly, for the critical value of β, the fixed point stability is altered and the trajectories shift away from the domain of the previous attractor (Fig. 15c,d). Additionaly, inside the area of this strange attractor there exists a space which does not contain any points. This space was previously covered by the points of the Poincaré map presented in Fig. 15a. For $\beta = 0.098$ we observe again the long chaotic transitional phenomena and then for $t > 1000T$ the points of the Poincaré maps lying on the closed curves. Because the corresponding frequency spectra are discrete (with two components) the corresponding attractor is quasiperiodic. With a very small increase of β (β=0.1) the chaotic transitional phenomena do not appear and a quasiperiodic attractor is clearly visible. As can be seen in Fig. 15f, we have two frequency components $\omega_1 = 12.3$, $\omega_2 = 16.7$. For $\beta > 0.1$ (more precisely $\beta = 0.11$,

$\beta = 0.12$) no changes have been found in the behaviour of the system (however the situation completly changes for $\delta = 0.11$- see Section 2.3.3.).

Comparison of these results with the "crisis phenomena" suggested by Grebogi and Ott [64] provides an interesting insight. The authors define a crisis as a collision between a chaotic attractor and a coexisting unstable fixed point. They distinguish two types of "crisis", the "boundary" and the "interior" crisis. The first leads to sudden destruction of the chaotic attractor and its basin of attraction, while the second can cause sudden changes in the size of the chaotic attractor. In our case however one can observe another phenomenon. First we have obtained a fixed point on the Poincaré map and then a very long chaotic transitional phenomenon appeared which for the further increase of β changes into a strange attractor. This provides the evidence for another new unique "crisis" type not analysed by Grebogi and Ott, where the changes of stability of the previously stable fixed point cause the shift to irregular motion. Now we can consider the new attractor of mixed type containing both a coexisting, chaotic attractor and an unstable fixed point. With further increase of β, another crisis occurs and leads to both the destruction of a strange attractor and creation of a quasiperiodic attractor.

We have analysed the evolution of Poincaré maps with the change of the nonlinear rigidity coefficient μ. The other parameters are: $\alpha = \beta = \delta = 0.1$, $\eta = 7.8$, $q = 1.0$. For $\mu = 40.0$ we have discovered regular motion with one frequency equal to η. With further increase in μ the transitional phenomenon becomes longer and longer. Increase of μ to the value of $\mu = 65.0$ results in chaos. Then for $\mu = 68.5$, a two frequency motion is observed.

Evolution of the motion with change in δ has also been analysed ($\alpha = \beta = 0.1$, $\mu = 69.17$, $\eta = 7.8$, $q = 1$). For $\delta = 0.12$ we have found the stable fixed point on the Poincaré map. With a slight decrease of δ ($\delta = 0.115$) the previous stable point became unstable and chaos appeared. Inside the area of the strange attractor there was a domain without points. This domain was previously covered by points of the attractor for $\delta = 0.12$. With further decrease of δ (for $\delta = 0.11$) we have obtained a "weak" chaotic attractor. Then for $\delta = 0.09$ the motion becomes regular and a quasiperiodic attractor with two frequencies appeared. For the results of the numerical analysis (for the parameters near point P_0) we have marked in Fig. 16 the regions of chaotic motion. It can be seen that chaotic orbits appear in the neighborhood of 2η, $\eta/2$, $3/2\eta$.

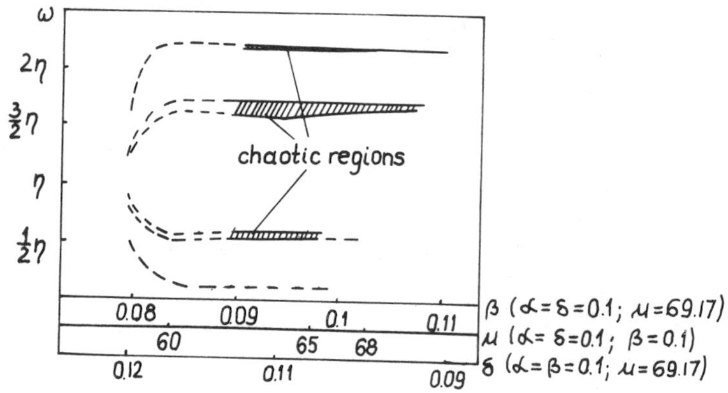

Fig. 16. Regions of chaotic motion for parameters near the point P_0.

Finally, we have investigated the route to chaos for parameters of the system placed near the bifurcation points R_0, S_0, T_0. For parameters near the point R_0 ($\alpha = 1.0$, $\beta = 0.1$, $\delta = 0.3$, $\eta = 1.6$, $q = 3$) the evolution from regular motion ($\mu = 14.3$) through chaos ($\mu = 14.35 \div 14.45$) to quasiperiodic motion ($\mu = 15.0$) was observed.

For parameters $\alpha = 1.0$, $\beta = 0.1$, $\delta = 0.3$, $\eta = 5.6$, $q = 3$ (near the point S_0) and $\mu = 135$ there are four frequencies in the motion, but the magnitude of the amplitudes corresponding to these frequencies evolve with time and the Poincaré map has a complicated structure. Frequency spectra are more broad-band and the corresponding amplitudes vary in time for $\mu =$

137. With a further increase of μ ($\mu = 139$) a quasiperiodic attractor appeared.

For a set of parameters near the point T_0 ($\alpha = 1.0$, $\beta = 0.1$ $\delta = 0.3$, $\eta = 10.0$, $q = 3.0$) we have observed transitional chaotic phenomena at $\mu = 440.0$, and then, for $\mu = 442.0$ and $\mu = 443.0$, appearance of the strange chaotic attractor. For $\mu = 446.0$ chaotically transitional phenomena occured for $t \leqslant 900T$. For $t > 900T$ we obtained the quasiperiodic attractor with three frequencies.

It is interesting that all strange chaotic attractors discovered possess a horizontal axis of symmetry.

3.2.2. Bifurcation of the quasiperiodic orbit with two frequencies

We assume that the stationary solution has the form

$$y = Y + A\cos\eta t + B\sin\eta t + R\cos\Omega t, \tag{89}$$

where Ω is the dimensionless frequency of self excited vibrations. From equations (81), taking into account (89), one can obtain the nonlinear algebraic equation set

$$Y(\alpha + \mu(Y^2 + \tfrac{3}{2}(P^2 + R^2))) - q = 0,$$

$$\beta - \delta(Y^2 + \tfrac{1}{2}P^2 + \tfrac{1}{4}R^2) = 0,$$

$$R(\alpha - \Omega^2) + 3\mu R(Y^2 + \tfrac{1}{2}P^2 + \tfrac{1}{4}R^2) = 0,$$

$$B(\alpha - \eta^2) - A\eta(-\beta + \delta(Y^2 + \tfrac{1}{4}P^2 + \tfrac{1}{2}R^2)) + 3\mu B(Y^2$$

$$+ \tfrac{1}{4}P^2 + \tfrac{1}{2}R^2) = 0, \tag{90}$$

$$A(\alpha - \eta^2) + B\eta(-\beta + \delta(Y^2 + \tfrac{1}{4}P^2 + \tfrac{1}{2}R^2)) + 3\mu A(Y^2$$

$$+ \tfrac{1}{4}P^2 + \tfrac{1}{2}R^2) - \eta^2 = 0,$$

$$P^2 = A^2 + B^2.$$

Consider now the perturbation equation

$$y_p = Y + \Delta Y + (A + \Delta A)\cos\eta t + (B + \Delta B)\sin\eta t$$

$$+ (R + \Delta R)\cos\Omega t + \Delta x\sin\Omega t. \tag{91}$$

Assuming that the values of the coefficients β, δ, Δ· are small from equation (91) we obtain

$$-2\eta\Delta\dot{A} + c_{11}\Delta A + c_{12}\Delta B + c_{13}\Delta R + c_{14}\Delta x = 0,$$

$$2\eta\Delta\dot{B} + c_{21}\Delta A + c_{22}\Delta B + c_{23}\Delta R + c_{24}\Delta x = 0, \tag{92}$$

$$-2\Omega\Delta\dot{R} + c_{31}\Delta A + c_{32}\Delta B + c_{33}\Delta R + c_{34}\Delta x = 0,$$

$$2\Omega\Delta\dot{x} + c_{41}\Delta A + c_{42}\Delta B + c_{43}\Delta R + c_{44}\Delta x = 0,$$

where

$$c_{11} = -\eta(-\beta + \delta(Y^2 + \tfrac{1}{4}B^2 + \tfrac{1}{2}R^2)) - \tfrac{3}{4}\delta\eta A^2 + \tfrac{3}{2}\mu AB,$$

$$c_{12} = \alpha - \eta^2 + 3\mu(Y^2 + \tfrac{1}{4}A^2 + \tfrac{1}{2}R^2) + \tfrac{9}{4}\mu B^2 - \tfrac{1}{2}\delta\eta AB,$$

$$c_{13} = -\delta\eta AR + 3\mu BR,$$

$$c_{14} = 0,$$

$$c_{21} = \alpha - \eta^2 + 3\mu(Y^2 + \tfrac{1}{4}B^2 + \tfrac{1}{2}R^2) + \tfrac{9}{4}\mu A^2 + \tfrac{1}{2}\eta\delta AB,$$

$$c_{22} = \eta(-\beta + \delta(Y^2 + \tfrac{1}{4}A^2 + \tfrac{1}{2}R^2)) + \tfrac{3}{4}\delta\eta B^2 + \tfrac{3}{2}\mu AB,$$

$$c_{23} = 3\mu AR + \eta\delta BR,$$

$$c_{24} = 0, \tag{93}$$

$$c_{31} = -\delta A,$$

$$c_{32} = -\delta B,$$

$$c_{33} = -\tfrac{1}{2}\delta R,$$

$$c_{34} = \alpha - \Omega^2 + 3(Y^2 + \tfrac{1}{2}(P^2 + R^2)),$$

$$c_{41} = 3\mu RA,$$

$$c_{42} = 3\mu RB,$$

$$c_{43} = \alpha - \Omega^2 + 3\mu(Y^2 + \tfrac{1}{2}P^2 + \tfrac{3}{4}R^2),$$

$$c_{44} = (-\beta + \delta(Y^2 + \tfrac{1}{2}P^2 - \tfrac{1}{4}R^2))\Omega .$$

The eigenvalues of (93) we obtain from

$$16\eta^2\Omega^2\lambda^4 + 8\Omega\eta((c_{44} - c_{33})\eta + \Omega(c_{22} - c_{11}))\lambda^3 +$$

$$+ 4(\eta\Omega(c_{11} - c_{22})c_{33} + \Omega^2(c_{12}c_{21} - c_{11}c_{22})$$

$$+ \eta\Omega(c_{32}c_{23} - c_{13}c_{31}) + \eta^2 c_{43}c_{34} - \eta c_{44}(\eta c_{33}$$

$$+ \Omega(c_{11} - c_{22})))\lambda^2 + 2(\Omega(c_{13}c_{21}c_{32} + c_{12}c_{23}c_{31}$$

$$- c_{13}c_{22}c_{31} + c_{11}c_{22}c_{33} - c_{12}c_{21}c_{33} - c_{32}c_{23}c_{11})$$

$$+ \eta c_{34}(c_{22}c_{43} + c_{13}c_{41} - c_{11}c_{43} - c_{23}c_{42}) + c_{44}(\eta$$

$$(c_{11} - c_{22})c_{33} + \Omega(c_{12}c_{21} - c_{11}c_{22}) + \eta(c_{32}c_{23}$$

$$- c_{13}c_{31})))\lambda + c_{34}(c_{12}c_{21}c_{43} + c_{11}c_{23}c_{42} + c_{13}c_{22}c_{41}$$

$$- c_{13}c_{21}c_{42} - c_{12}c_{23}c_{41} - c_{11}c_{22}c_{43}) + c_{44}(c_{13}c_{21}c_{32}$$

$$+ c_{12}c_{23}c_{31} + c_{11}c_{22}c_{33} - c_{12}c_{21}c_{33} - c_{32}c_{23}c_{11}$$

$$- c_{13}c_{22}c_{31}) = 0. \qquad (94)$$

The necessary conditions of the existence of a Hopf bifurcation point is the existence, in equation (94), of two purely imaginary eigenvalues, whereas all other eigenvalues have negative real parts. The full system of bifurcation equations in this case are obtained by assigning the expresions for λ^3, λ^2 and λ to zero while the last expression ought to be greater than zero. Unfortunately, using this technique we are unable to find bifurcation curves, but we have found two isolated solutions. In the first case the critical parameters are: $\alpha = 0.05$, $\beta = 0.019$, $q = 0.00253$, $\mu = 6.274$, $\eta = 0.1218$, $\delta = 0.4$. Hopf type bifurcation takes place by changing the δ parameter. From numerical experiments we have obtained for δ = 0.4 a periodic motion and, with further increase of this coefficient, we can observe many bifurcations which lead to subharmonic and quasiperiodic motion. In this case we have not detected chaotic motion.

The second example is presented in Fig. 17. For δ = 0.6 one

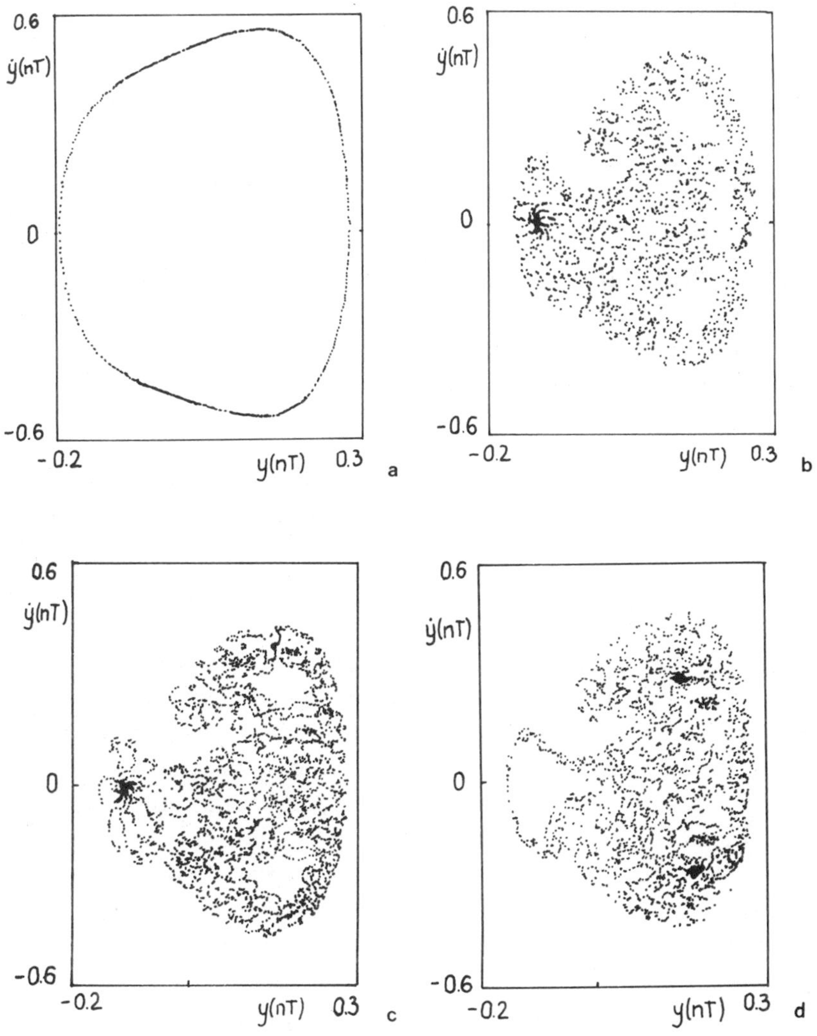

Fig. 17.

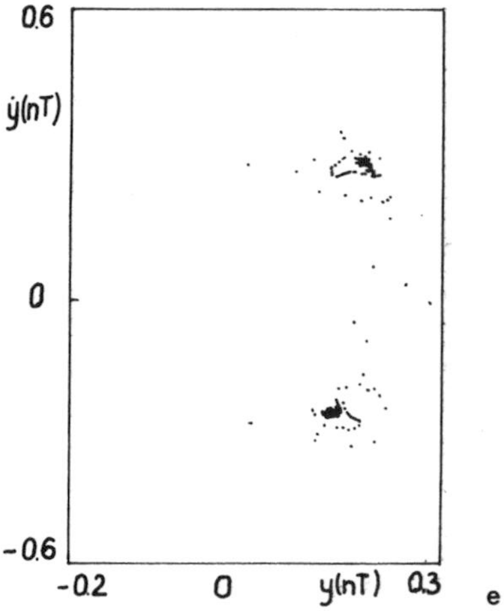

Fig. 17. Evolution of attractors for the two - frequency solution: $\alpha = \beta = 0.02$; $q = -0.2665$; $\mu = 100.1$; $\eta = 0.724$; a) $\delta = 0.6$; b) $\delta = 0.73$; c) $\delta = 0.76$; d) $\delta = 0.762$; e) $\delta = 0.9$.

can observe quasiperiodic motion and, for $0.6 < \delta < 0.7$ Hopf bifurcation of the stationary state occurs. We have presented in Fig. 17b the chaotic transitional phenomena which persist to 1400 periods and then converge to one point (the map has 2000 points). For $\delta = 0.76$ the situation is similar to that previously described but suddenly, for $\delta = 0.762$ (see Fig. 15d), the stable fixed point vanishes and two new, stable symmetric points appear. It is strongly evident that the new, stable fixed point appears in the middle of the domain previously free of points and, further, that new domains free of points inside the attractor are located at the same place where, earlier, the single fixed point was lying. Generally the shape of attractor remains unchanged during changes of δ.

Therefore, using the example of this simple anharmonic oscillator, we have detected a new phenomenon indicative of the simultaneous coexistence of one stable and two unstable fixed points with long transitional chaos. For $0.76 < \delta < 0.762$

bifurcation appears so that the stability for these three points is changed with the previous one stable point becoming unstable and the two unstable points becoming stable. However, this changing of stability of the fixed points appears to have no influence on the transitional chaotic phenomena because the shape of the Poincaré map remains unchanged. With further increase of δ the chaotic transitional phenomena become shorter and shorter. During these changes, however, the two stable fixed points remain at their previous positions. Eventually, the transitional chaotic behaviour disappears (Fig. 17e) and we find two stable fixed points. The changes in the behaviour of chaotic transitional phenomena have no apparent influence on the other stable fixed points and vice versa.

3.2.3. Summary and concluding remarks

Using an approximate analytical method, we have obtained a set of bifurcation equations, assuming that bifurcation of the stationary regular motion with one and two frequencies takes place. In the former case we have determined bifurcation curves, while in the latter we have found only two isolated solutions.

In the first case we have presented some examples of strange chaotic attractors. All of them possess a horizontal axis of symmetry even though the static load in the considered equation was expected to cause nonsymmetry.

In Fig. 15 we have shown, with the changes of stability of the previously stable fixed point, how the long chaotic transitional phenomena shifts into a chaotic attractor. This new complex attractor can be considered as a mixed one in which a strange chaotic attractor and an unstable fixed point coexist.

We have described a route from a periodic motion to an irregular one with the change of the nonlinear rigidity μ. With the increase of μ chaotically transitional phenomena last longer and then a strange chaotic attractor appears.

The analog scheme of transition from regular to irregular motion is also presented when we have discovered chaos by changing the parameter δ. Similar as in two earlier cases, inside the domain of the strange attractor there is a region without points. This space was previously covered by the points of the stable attractor, which now is unstable and coexists with the strange chaotic attractor.

With reference to bifurcation of the stationary state with two frequencies, two isolated critical values of the parameters were found. For the first set many bifurcations appear with a variation of the coefficient δ. However, this does not lead to chaos. Near the second set of bifurcation parameters analysed, the evolution of simultaneously coexisting stable and unstable fixed points with chaotically transitional phenomena (Fig. 15) is detected. This evolution shows that fixed points and these transitional phenomena are "uncoupled" in the sense that changes in one behaviour do not have any influence on the other. In our investigations we have used the discrete power spectrum. It has allowed us to observe the evolution of the Fourier components with time.

3.3. Particular Van der Pol-Duffing's Oscillator

3.3.1. The analysed system and averaged equations

The equation of motion for the considered system is

$$M\ddot{x} + (c_2x^2 - c_1)\dot{x} + k_0x + k_1x^3 + \mu_0 Mg\,\mathrm{sgn}\dot{x} = P_0(t) \qquad (95)$$

where $P_0(t) = a\mu m\nu^2\cos\nu t$.

The exciting force $P_0(t)$ originates from the rotating engine rotor with the mass m and the unbalance μ, where a is the amplification coefficient. From (95) we obtain

$$\dot{y} = z,$$

$$\dot{z} = \varepsilon(1 - y^2)\dot{y} - \delta y - \gamma y^3 - \alpha\,\mathrm{sgn}\dot{y} + p_0\nu^2\cos\nu t, \qquad (96)$$

where:

$$y = (c_2/c_1)^{1/2}x, \quad \varepsilon = c_1/M, \quad \delta = k_0/M, \quad \gamma = k_1c_1/Mc_2,$$

$$= \mu_0 g(c_2/c_1)^{1/2}, \quad p_0 = a\mu m/M(c_2/c_1)^{-1/2}. \qquad (97)$$

The approximate analytical method is used to solve the system of equations (96) assuming that

$$\begin{aligned} y &= u(t)\cos\nu t - v(t)\sin\nu t, \\ z &= -\nu(u(t)\sin\nu t + v(t)\cos\nu t), \end{aligned} \qquad (98)$$

where u(t) and v(t) are the slowly changing functions of t. After substituting equations (98) in (96) we obtain:

$$\begin{aligned} \dot{u} &= N\sin\nu t, \\ \dot{v} &= N\cos\nu t, \end{aligned} \qquad (99)$$

where:

$$\begin{aligned} N = {} & (\delta - \nu^2)\nu^{-1}(u\cos\nu t - v\sin\nu t) + \varepsilon\nu^{-1}(u\cos\nu t \\ & - v\sin\nu t)^3 + \varepsilon(1 - (u\cos\nu t - v\sin\nu t)^2)(u\sin\nu t \\ & + v\cos\nu t) + \alpha/\nu \,\mathrm{sgn}(-\nu(u\sin\nu t + v\cos\nu t)) \\ & - p_0\nu\cos\nu t. \end{aligned} \qquad (100)$$

The right sides of equation (99), being periodic functions of t with the period $2\pi/\nu$, are expanded into Fourier series. Taking into consideration that both u and v are the slowly changing functions of t and that only the first terms of expansions are significant the following averaged equations system is obtained:

$$\dot{u} = -\frac{\delta-\nu^2}{2}v + \frac{\varepsilon}{2}u - \frac{\varepsilon}{2}u(u^2+v^2) - \frac{3\gamma}{8\nu}v(u^2+v^2)$$

$$-\frac{2\alpha}{\pi\nu}u(u^2+v^2)^{-1/2},$$

$$\dot{v} = \frac{\delta-\nu^2}{2\nu}u + \frac{\varepsilon}{2}v - \frac{\varepsilon}{8}v(v^2+u^2) + \frac{3\gamma}{8\nu}u(u^2+v^2)$$

$$-\frac{2\alpha}{\pi\nu}v(u^2+v^2)^{-1/2} - \frac{p_0\nu}{2}. \qquad (101)$$

Assuming that

$$\omega = \frac{\delta-\nu^2}{\nu}, \quad \varepsilon = 8, \quad \frac{3\gamma}{8\nu} = 1, \quad \frac{2\alpha}{\pi\nu} = 1, \quad \frac{\nu}{2} = P,$$

$$p_0 = 1, \qquad (102)$$

we obtain

$$\dot{u} = -\omega v + 4u - u(u^2+v^2) - v(u^2+v^2)$$

$$-u(u^2+v^2)^{-1/2},$$

$$\dot{v} = \omega u + 4v - v(u^2+v^2) + u(u^2+v^2)$$

$$-v(u^2+v^2)^{-1/2} - P. \qquad (103)$$

Using the polar coordinates (r, Θ) we have

$$u = r\cos\Theta,$$
$$v = r\sin\Theta, \tag{104}$$

and the equations system (103) acquires the form

$$\dot{r} = -r^3 + 4r - 1 - P\sin\Theta,$$
$$\dot{\Theta} = r^2 + \omega - \frac{P}{r}\cos\Theta. \tag{105}$$

At singular points $\dot{r} = \dot{\Theta} = 0$, and from equations (105) we get

$$r^6 + (\omega - 4)r^4 + r^3 + \frac{\omega^2+16}{2} r^2 - 4r + \frac{1-P^2}{2} = 0. \tag{106}$$

The use of the approximation method, known as Van der Pol method, is very convenient, particularly in order to obtain the bifurcations of an averaged system of the forced oscillators. Only one assumption must be taken into account, that the amplitudes u(t) and v(t) change slowly in time t. The same method was used, for instance, by Arrowsmith and Taha [63], where the local and global bifurcations of the averaged two parameter system were investigated (see also [64]).

3.3.2. "0"-type bifurcations

The "0" type bifurcations occur when singular points overlap. The necessary condition for the existence of multiple roots of the equation (106) has the form:

$$a_0^{10}\begin{vmatrix} s_0 & s_1 & s_2 & s_3 & s_4 & s_5 \\ s_1 & s_2 & s_3 & s_4 & s_5 & s_6 \\ s_2 & s_3 & s_4 & s_5 & s_6 & s_7 \\ s_3 & s_4 & s_5 & s_6 & s_7 & s_8 \\ s_4 & s_5 & s_6 & s_7 & s_8 & s_9 \\ s_5 & s_6 & s_7 & s_8 & s_9 & s_{10} \end{vmatrix} = 0, \qquad (107)$$

where:

$$s_k = (-\frac{1}{a_0})^k \begin{vmatrix} a_1 & a_0 & 0 & \cdots\cdots & 0 \\ 2a_2 & a_1 & a_0 & \cdots\cdots & 0 \\ 3a_3 & a_2 & a_1 & \cdots\cdots & 0 \\ \cdot & \cdot & \cdot & & \\ \cdot & \cdot & \cdot & & \\ ka_k & a_{k-1} & & & a_1 \end{vmatrix} \qquad (108)$$

$$a_0 = 1, \quad a_4 = \frac{\omega^2+16}{2}, \quad a_8 = 0,$$

$$a_1 = 0, \quad a_5 = -4, \quad a_9 = 0,$$

$$a_2 = \omega - 4, \quad a_6 = \frac{1-P^2}{2}, \quad a_{10} = 0,$$

$$a_3 = 1, \quad a_7 = 0, \quad s_0 = 6.$$

From equation (108) we obtain:

$$x^5 + b_4x^4 + b_3x^3 + b_2x^2 + b_1x + b_0 = 0, \qquad (109)$$

where:

$$x = P^2,$$

$$b_4 = 0.7\omega^3 + 16\omega^2 - 71.1\omega - 1.1,$$

$$b_3 = 0.2\omega^6 + 21\omega^5 - 28\omega^4 - 360.1\omega^3 + 704.2\omega^2 + 5244.2\omega - 2412.4,$$

$$b_2 = -0.2\omega^9 + 3\omega^8 + 18.5\omega^7 - 336.4\omega^6 - 1440\omega^5 + 36330\omega^4 + 1388.5\omega^3 - 7608.7\omega^2 - 19960\omega + 62152,$$

$$b_1 = 0.001\omega^{12} + 0.6\omega^{11} - 0.07\omega^{10} - 125.2\omega^9 - 623.4\omega^8 + 9400.9\omega^7 + 30163\omega^6 - 26595\omega^5 + 427720\omega^4 + 93038\omega^3 + 333350\omega^2 + 3347200\omega + 4174400, \quad (110)$$

$$b_0 = -0.8\omega^{13} - 14.5\omega^{12} - 95.5\omega^{11} - 625.1\omega^{10} + - 25696.5\omega^9 - 41611.7\omega^8 + 1058200\omega^7 - 419820\omega^6 + - 5740700\omega^5 - 378800\omega^4 + 5201600\omega^3 + 1165600\omega^2 + 5370900\omega + 977300.$$

Considering the character of the equation (109), the dependence $P(\omega)$ has been determined numerically by means of substituting the value $\omega \in \mathbb{R}$ and then solving the algebraic equation in terms of x using the secant method.

3.3.3. Bifurcation of the periodic orbit

In order to determine the equation for the complex bifurcations, the terms of the equation (101), containing the

square root, are expanded into the Taylor series around the temporarily unknown point of equilibrium (u_0, v_0). The expansion is limited to the linear terms of u and v. As a result, from equations (101), we have:

$$\dot{u} = -\omega v + (4 - (u^2 + v^2))u - (u^2 + v^2)v - \frac{u_0}{\sqrt{u_0^2 + v_0^2}}$$

$$- \frac{v_0^2}{(u_0^2 + v_0^2)^{3/2}} (u - u_0) + \frac{u_0 v_0}{(u_0^2 + v_0^2)^{3/2}} (v - v_0),$$

$$\dot{v} = \omega u + (4 - (u^2 + v^2))v + (u^2 + v^2)u$$

$$- \frac{v_0}{\sqrt{u_0^2 + v_0^2}} - \frac{u_0^2}{(u_0^2 + v_0^2)^{3/2}} (v - v_0)$$

$$+ \frac{u_0 v_0}{(u_0^2 + v_0^2)^{3/2}} (u - u_0) - P. \qquad (111)$$

Let us now proceed to the new coordinate system (u', v') whose origin is the singular point (u_0, v_0). From equations (111), after linearization, we obtain

$$\dot{u}' = (4 - 3u_0^2 - v_0^2 - 2u_0 v_0 - \frac{v_0^2}{(u_0^2 + v_0^2)^{3/2}})u'$$

$$+ (-\omega - u_0 - 3v_0^2 - 2u_0 v_0 + \frac{u_0 v_0}{(u_0^2 + v_0^2)^{3/2}})\, v',$$

$$\dot{v}' = (\omega + 3u_0^2 + v_0^2 - 2u_0 v_0 + \frac{u_0 v_0}{(u_0^2 + v_0^2)^{3/2}})\, u'$$

$$+ (4 - u_0^2 - 3v_0^2 + 2u_0 v_0 - \frac{u_0^2}{(u_0^2 + v_0^2)^{3/2}})\, v'. \qquad (112)$$

The characteristic equation of the system equations (112) has the form

$$\delta^2 - (A + D)\delta + AD - BC = 0, \tag{113}$$

where

$$A = 4 - 3u_0^2 - v_0^2 - 2u_0v_0 - \frac{v_0^2}{(u_0^2 + v_0^2)^{3/2}},$$

$$B = -\omega - u_0^2 - 3v_0^2 - 2u_0v_0 + \frac{u_0v_0}{(u_0^2 + v_0^2)^{3/2}},$$

(114)

$$C = \omega + 3u_0^2 + v_0^2 - 2u_0v_0 + \frac{u_0v_0}{(u_0^2 + v_0^2)^{3/2}},$$

$$D = 4 - u_0^2 - 3v_0^2 + 2u_0v_0 - \frac{u_0^2}{(u_0^2 + v_0^2)^{3/2}}.$$

In order for the Hopf bifurcation to exist, it is necessary for the roots of the characteristic equation (113) to be strictly imaginary. This leads to the condition that

$$A + D = 0,$$

(115)

$$AD - BC > 0.$$

From the first equation of the system (115) and from equations (101) we obtain the set of bifurcation equations

$$\frac{1}{4\sqrt{u_0^2 + v_0^2}} = 2 - (u_0^2 + v_0^2),$$

$$-\omega v_0 + (4 - (u_0^2 + v_0^2))\, u_0 - (u_0^2 + v_0^2)\, v_0$$

$$- \frac{u_0}{\sqrt{u_0^2 + v_0^2}} = 0, \tag{116}$$

$$\omega u_0 + (4 - (u_0^2 + v_0^2))v_0 + (u_0^2 + v_0^2)u_0 +$$

$$- \frac{v_0}{\sqrt{(u_0^2 + v_0^2)}} - P = 0.$$

This equation set, after replacing the carthesian coordinates (u_0, v_0) by polar ones (r_0, Θ_0) acquires the following form

$$4r_0(2 - r_0^2) = 1,$$

$$- \omega r_0 \sin\Theta_0 + (4 - r_0^2)\, r_0 \cos\Theta_0 - r_0^3 \sin\Theta_0 - \cos\Theta_0 = 0,$$

$$\omega r_0 \cos\Theta_0 + (4 - r_0^2)\, r_0 \sin\Theta_0 + r_0^3 \cos\Theta_0 - \sin\Theta_0$$

$$- P = 0. \tag{117}$$

From the first equation of (117) three values of r_0 are obtained: $r_{01} = 1.347$, $r_{02} = 0.126$ and $r_{03} = - 1.473$.

After substituting these values into the other two equations in (117), three necessary conditions for the Hopf bifurcation are obtained

$$P^2 = 1.814\omega^2 + 6.584\omega + 2.274,$$

$$P^2 = 0.016\omega^2 + 0.001\omega + 2.01, \tag{118}$$

$$P^2 = 2.17\omega^2 + 9.415\omega - 34.409.$$

Figure 18 presents the complex (Hopf) bifurcation curves along with the diagram of the "O" type bifurcation curves. With regard to the symmetry of the equations in relation to P, the figure only presents the half plane $P \geqslant 0$.

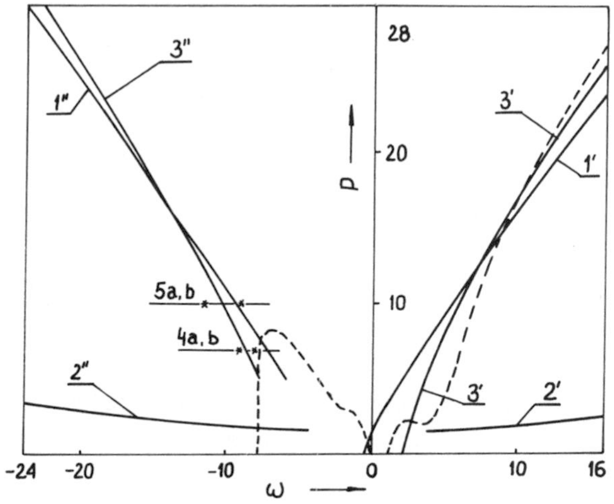

Fig. 18. Hopf (———) and the "O" type (- - -) bifurcation curves.

3.3.4. Observations of strange chaotic attractors

As many authors connect the occurrence of chaos with bifurcations, it seems appropriate to seek irregular motion on the plane with parameters P, ω near the bifurcation curves (Fig. 18). For this purpose the differential equation (96) has been numerically solved, after taking into account the dependencies (102). Numerical calculations were made by using the variable-order, variable-step Gear method. The unknown solution is obtained by interpolation on solution values produced by the Gear method. The accuracy of the integration and the interpolation is controlled by two parameters. The first is a calculation step with which integrations are made, while the latter determines the type of error control. At each step in the numerical solution an estimate of the local error is made. If the appropriate condition is not satisfied then the step-size is reduced and the solution is recomputed on the current step. The Gear method was used in order to avoid problems with numerical accuracy, when dealing with such an ill behaved nonlinearity as dry friction. A description of Gear methods and their practical implementation is given in [65].

Numerical results are presented as time histories, phase portraits and Poincaré maps. The first digital integrations run for a long time (about 300 s) so that all transients have decayed, which allows the trajectory to reach the final attractor. Then they are recorded to the time 1600T ($T = 2\pi/\nu$) in order to obtain Poincaré maps. In all presented figures $Y_1 = y$, $Y_2 = z$.

The introduction of the connections (102) between parameters of the nonlinear oscillator (96) makes it possible to observe the behaviour of this oscillator on the two parameter plane when bifurcation appears (see Fig. 18). Generally, chaotic motion has been found for the parameters lying near the bifurcation curves. In addition, with increasing ω for the established values of P, a tendency towards ordering of the motion is observed. Previously Grebogi and Ott [64] had discussed the occurrence of sudden qualitative changes of chaotic dynamics as a single parameter is varied. Sudden changes in the size, sudden appearences and sudden destruction of chaotic attractors have been connected with the collision crises of an unstable periodic orbit and a coexisting chaotic attractor. In our example, we encounter a similar situation in the sense that, for ω lying sufficiently left of the bifurcation curves, we have discovered the regular orbits. With increasing ω, for the parameters near the bifurcation points, the orbits fail to be closed curves and strange chaotic attractors were detected. Earlier, Curry and Yorke [66] have shown by the example of the particular maps in R^2 that at certain parameter values a Hopf bifurcation occurs, and as parameters change, the attracting invariant circle grows and starts to warp, eventually becoming a strange attractor.

We should underline in this place the benefit of using the approximate method described earlier. Thanks to averaged equations (99) obtained by this method, the Hopf bifurcation is related to the bifurcations of the amplitudes u and v of the periodic orbits in equations (96). It means that the conditions for Hopf bifurcations are related to the bifurcation of periodic orbits in the preliminary equations (96).

We show only some of the examples of chaotic motion detected near the bifurcation curves. We will discuss and illustrate the behaviour of the nonlinear oscillator with the increase in ω for the established values of P (see marked horizontal lines in Figure 18). For $P = 7$ and $\omega = -9$ we have detected a chaotic strange attractor. The time trajectory, phase orbit and Poincaré map prove this. With further increase of ω the strange attractor changes but remains chaotic (Fig. 20b).

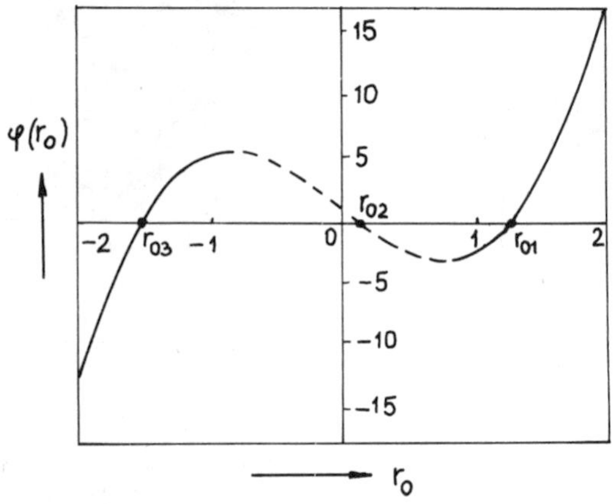

Fig. 19. The equilibrium points corresponding to the Hopf bifurcation curves presented in Figure 18.

Consider now the behaviour of the system in a similar way but for larger values of P. The strange chaotic attractor is presented in Fig. 21a for $P = 10$ and $\omega = -11.5$. With increase of ω, the chaotic dynamics of the orbits become weak (Fig. 21b).

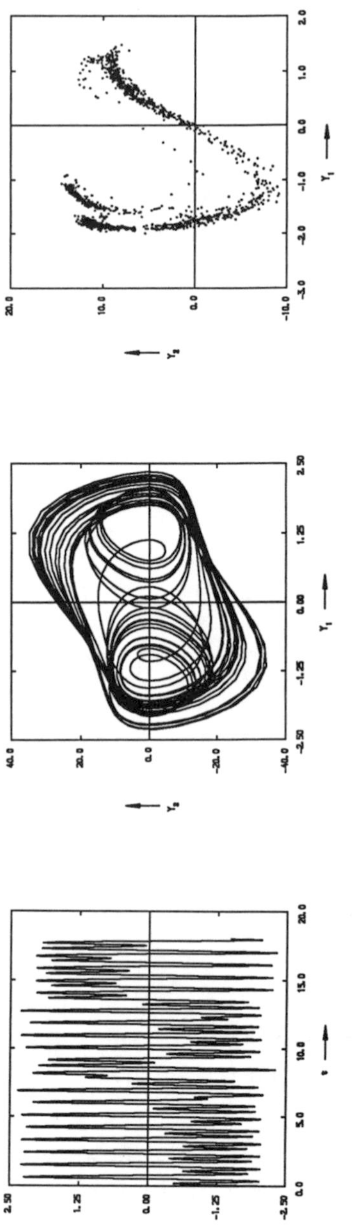

20.0
10.0
0.0
-10.0
-3.0
-2.0
-1.0
0.0
1.0
2.0
40.0
20.0
0.0
-20.0
-40.0
-2.50
-1.25
0.00
1.25
2.50
2.50
1.25
0.00
-1.25
-2.50
0.0
5.0
10.0
15.0
20.0

20a

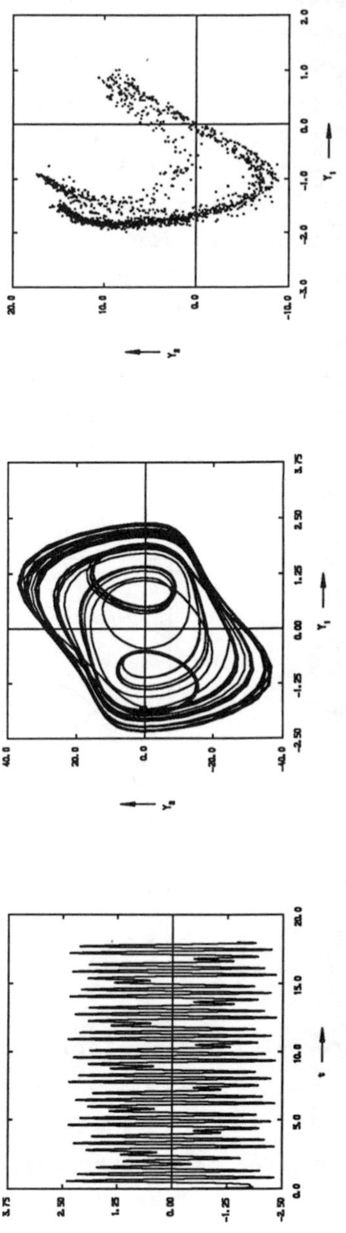

20.0
10.0
0.0
-10.0
-3.0
-2.0
-1.0
0.0
1.0
2.0
40.0
20.0
0.0
-20.0
-40.0
-2.50
-1.25
0.00
1.25
2.50
3.75
3.75
2.50
1.25
0.00
-1.25
-2.50
0.0
5.0
10.0
15.0
20.0

20b

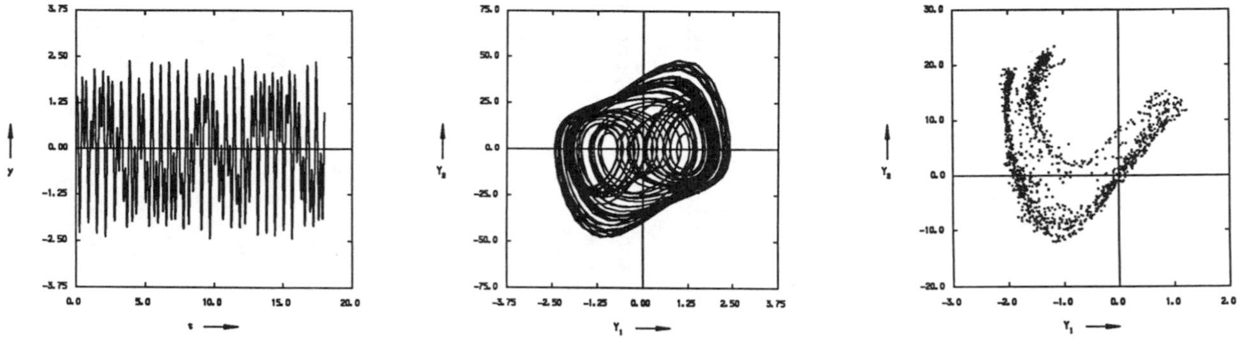

21a

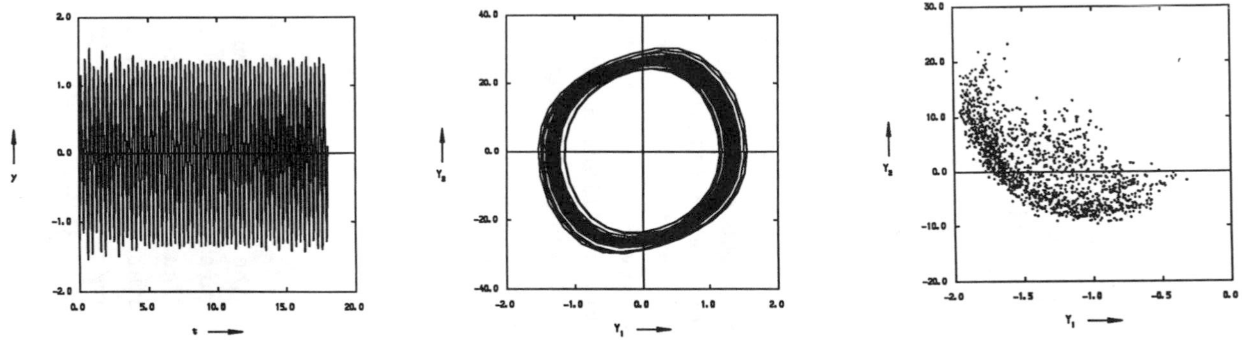

21b

Fig. 20, 21. Time histories, phase portraits and Poincaré maps for different sets of bifurcation parameters (ω, P) from Table 6.

Table 6. Parameters of equation (96).

FIGURE	POINT(ω;P)	$\varepsilon = 8.$; $P_0 = 1.$			
		δ	γ	α	ν
20a	(-9.;7.)	-56.	37.33	21.99	14.
20b	(-8.;7.)	-28.	37.33	21.99	14.
21a	(-11.5;10)	-60.	53.33	31.42	20.
21b	(-9.;10.)	40.	53.33	31.42	20.

During computer experiments, chaos has not been found near the bifurcation curves 2′ and 2″ . Figure 19 presents the diagram of dependencies $f(r_0)$, where the roots of the first equation of system (116) are marked. It is evident that the equilibrium point, r_{02}, is unstable. For this reason, the bifurcation curves marked in Fig. 18 as 2′ and 2″ correspond to the unstable equilibrium point of averaged equations. For the parameters near these bifurcation curves we have detected periodic and quasiperiodic attractors. Furthermore, the numerical investigations have not proven the existence of any relationship between the occurrence of "0"-type bifurcations and the occurrence of chaos in the considered oscillator.

Finally, we consider how the behaviour of the strange attractor depends on dry friction. In order to investigate this problem we take all the same parameters as in the case presented in Fig. 20a excluding α, which will be changed. The results of the numerical simulations are presented as Poincaré maps in Figure 22. For $\alpha = 0$ the strange attractor is presented in Figure 22a. With the increase of α the strange attractor starts to contract (Fig. 22b, Fig. 20a, Fig. 22c). Further increase of damping introduced by friction cause the phase flow to become more and more contracted (Fig. 22d).

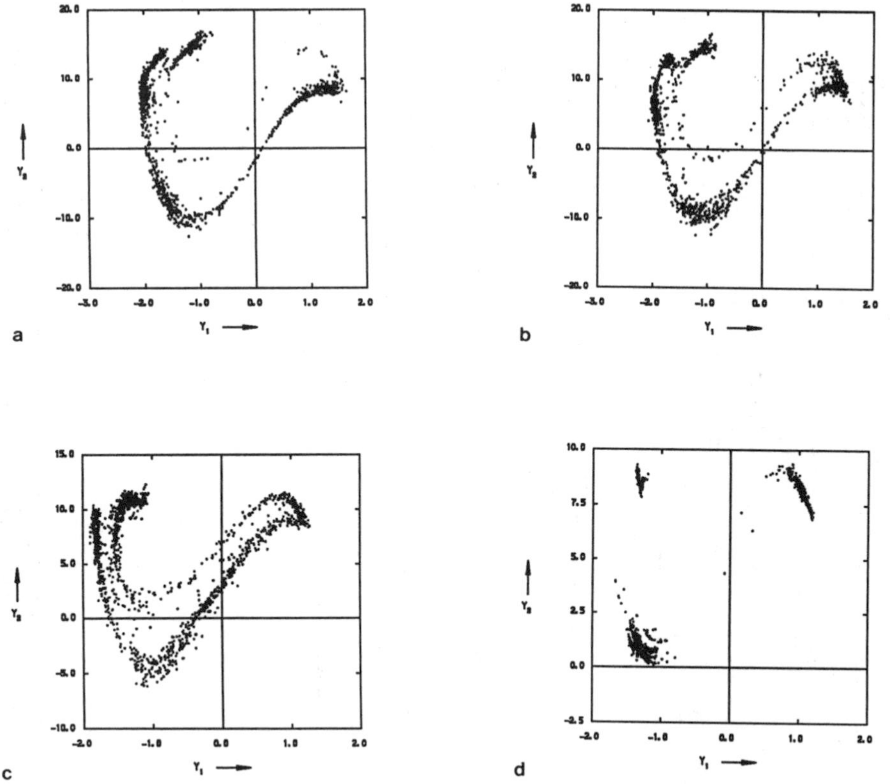

Fig. 22. The evolution of the strange attractor with change of the dry friction: a) $\alpha = 0$, b) $\alpha = 10$, c) $\alpha = 50$, d) $\alpha = 80$. The other parameters as in Figure 20.

3.4. Oscillator with Delay

3.4.1. Bifurcation of the periodic orbit

The equation of motion of the analysed system has the form

$$\ddot{x} - \alpha(1 - x^2)\dot{x} + \omega_0^2 x + \beta x^3 = kx(t - \tau_0) + F \cos \omega t. \tag{119}$$

It can, for instance, circumscribe the vibrations of the mechanical system presented in [67], where the linear spring force possess a time delay in its action. As we shall concentrate on the analysis of equation (119) for small τ_0 we have

$$x(t - \tau_0) = x(t) - \tau_0\dot{x}(t) + \frac{1}{2}\tau_0^2\ddot{x}(t). \tag{120}$$

From equation (119) we obtain

$$\ddot{x} - (p - \alpha_1 x^2)\dot{x} + qx + \beta_1 x^3 = F_1 \cos \omega t, \tag{121}$$

where:

$$\begin{aligned} p &= 2(\alpha - k\tau_0)(2 - k\tau_0^2)^{-1}, \\ \alpha_1 &= 2\alpha(2 - k\tau_0^2)^{-1}, \\ q &= 2(\omega_0^2 - k)(2 - k\tau_0^2)^{-1}, \\ \beta_1 &= 2\beta(2 - k\tau_0^2)^{-1}, \\ F_1 &= 2F(2 - k\tau_0^2)^{-1}. \end{aligned} \tag{122}$$

Equation (121) can be rewritten in the form

$$\begin{aligned} \dot{x} &= z, \\ \dot{z} &= (p - \alpha_1 x^2)\dot{x} - qx - \beta_1 x^3 + F_1 \cos \omega t. \end{aligned} \tag{123}$$

The approximate solution of (123) is forseen in the form:

$$\begin{aligned} x &= u \cos \omega t - v \sin \omega t, \\ z &= -\omega(u \sin \omega t + v \cos \omega t), \end{aligned} \tag{124}$$

where u and v are assumed to be slowly varying functions of t. From (124) we have

$$\dot{u} \cos \omega t - \dot{v} \sin \omega t = 0,$$

$$-\omega(\dot{u} \sin \omega t + v \cos \omega t) = \dot{z} + \omega^2 x, \qquad (125)$$

and, after solving for $\dot{u}$, $\dot{v}$

$$\dot{u} = -\frac{1}{\omega}(z + \omega^2 x) \sin \omega t,$$

$$\dot{v} = -\frac{1}{\omega}(\dot{z} + \omega^2 x) \cos \omega t, \qquad (126)$$

where

$$\dot{z} = \omega^2 x = (p - \alpha_1 x^2)\dot{x} + (\omega^2 - q)x - \beta_1 x^3 + F_1 \cos \omega t. \qquad (127)$$

From (126) follows after averaging

$$\dot{u} = \frac{q - \omega^2}{2\omega} v + \frac{pu}{2} - \frac{\alpha_1}{8} u(u^2 + v^2) - \frac{3\beta_1}{8} v(u^2 + v^2),$$

$$(128)$$

$$\dot{v} = \frac{q - \omega^2}{2\omega} u + \frac{pv}{2} - \frac{\alpha_1}{8} v(u^2 + v^2) + \frac{3\beta_1}{8} u(u^2 + v^2).$$

Let u_0 and v_0 be a steady state solution of (128), and let $u_1(t)$ and $v_1(t)$ be small disturbances of this solution. After substituting

$$u = u_0 + u_1,$$

$$(129)$$

$$v = v_0 + v_1,$$

in (128) and retaining only the terms of first powers of u_1 and v_1 we obtain

$$\dot{u}_1 = Au_1 + Bv_1,$$
$$\dot{v}_1 = Cu_1 + Dv_1, \tag{130}$$

where:

$$A = \frac{p}{2} - \frac{3}{8}\alpha_1 u_0^2 - \frac{\alpha_1}{8} v_0^2 - \frac{3\beta_1}{4\omega} u_0 v_0,$$

$$B = -\Omega - \frac{\alpha_1}{4} u_0 v_0 - \frac{9}{8}\frac{\beta_1}{\omega} v_0^2 - \frac{3}{8}\frac{\beta_1}{\omega} u_0^2,$$

$$C = \Omega - \frac{\alpha_1}{4} u_0 v_0 + \frac{9}{8}\frac{\beta_1}{\omega} u_0^2 + \frac{3}{8}\frac{\beta_1}{\omega} v_0^2, \tag{131}$$

$$D = \frac{p}{2} - \frac{\alpha_1}{8} u_0^2 - \frac{3}{8}\alpha_1 v_0^2 + \frac{3}{4}\frac{\beta_1}{\omega} u_0 v_0,$$

$$\Omega = \frac{q - \omega^2}{2\omega}.$$

The necessary conditions for the Hopf bifurcation are:

$$A + D = 0,$$
$$AD - CB > 0. \tag{132}$$

The first equation of (132) gives

$$\frac{2p}{\alpha_1} = u_0^2 + v_0^2, \tag{133}$$

which is satisfied when $\alpha_1 > k\tau_0$.

Now we introduce polar coordinates $r_0 \in (-\infty, +\infty)$ and $\Theta_0 \in (0, 2\pi)$. Then

$$u_0 = r_0 \cos\Theta_0, \qquad v_0 = r_0 \sin\Theta_0. \tag{134}$$

Because u_0 and v_0 satisfy (128), we obtain after substituting (134) into (128)

$$4\,\omega^2 r_0^2 \left[\frac{1}{4}(p - \frac{\alpha_1}{4} r_0^2)^2 + (\Omega + \frac{3}{8}\frac{\beta_1}{\omega} r_0^2)^2 \right] = F_1^2. \tag{135}$$

Introducing (133) in (135) gives

$$\frac{1}{2} p^3(\alpha_1\omega^2 + \frac{9\beta_1^2}{\alpha_1}) + 12\beta_1\omega\Omega p^2 + 8p\alpha_1\omega^2\Omega^2 - F_1^2\alpha_1^2 = 0. \tag{136}$$

While from the second condition of equation (132) we obtain

$$-\frac{1}{16} p^2 + \frac{27}{16}\frac{\beta_1^2 p^2}{\omega^2\alpha_1^2} p + \Omega^2 > 0. \tag{137}$$

The parameters of equation (121) should satisfy equation (136) and (137) in the critical point. When, additional, the real part of the complex conjugate eigenvalues of equation (130) are negative for $\alpha < \alpha_c$ and with the increase of α ($\alpha > \alpha_c$) their real part becomes positive a Hopf bifurcation occurs. In the considered case the critical value is α_c = 0.09 (the other parameter values are: $F = 109.54$, $\omega = 1.6$, $\omega_0 = 1.0$, $\beta = 10.0$, $k = 5.0$, $\tau_0 = 0.07$).

The numerical simulation of the equation (119) is made using the Runge-Kutta method, while the results are presented in the form of the time histories $x(t)$, phase portraits $\dot{x}(x)$, Poincaré maps and frequency spectra. Only the analysis of all four diagrams gives a full view of the behaviour of the system. Poincaré maps are registered after $T_{min} = 50$. It resulted from the performed numerical calculations that the duration of the transient state, caused by the used initial function is extended with the increase of delay τ_0, hence for small τ_0 a relatively small value of T_{min} has been assumed. The following initial function is introduced: $x(t) = 1.0$ for $-\tau_0 \leqslant t < 0$ and $x(t) = 0$ for $t = 0$, as well as $\dot{x}(t) = 0$ for $-\tau_0 \leqslant t \leqslant 0$.

The results of the numerical simulations are presented in Fig. 23. The calculations are performed with the integration step 0.01 for arbitrarily chosen parameters $F = 109.54$, $\omega = 1.6$ $\omega_0 = 1.0$, $\beta = 10.0$, $k = 5.0$, $\tau_0 = 0.07$ and Poincaré maps are made for $T_{max} = 1400$. One can observe the qualitative changes of the solution before the chaotic regime is reached. For a great damping coefficient $\alpha = 8$, the vibrations are periodic but with eight amplitudes of the Fourier components against frequency, which decrease almost exponentially (Fig. 24). With the decrease of the value of α (see Fig. 23b,c and corresponding Fourier spectra) the components of the solution with the other harmonics increase compared with the basic one. Moreover, the Fourier spectrum becomes more and more irregular. At $\alpha = 0.1$ new components in the frequency spectra appear and the spectrum in some narrow ranges looks like broad band – the motion exhibits small chaos (see Figs. 23d and 24d). With further decrease of α chaos becomes more profound. The strange attractors are presented on Poincaré maps (Fig. 23e,f,g) and the adequate frequency spectra are shown in Fig. 24e,f,g. The investigated equation, generally, is particularly sensitive to the changes of the time delay value τ_0. The increase of this value causes also an increase of the magnitude of the strange attractor (Fourier spectra become in this case more broad band), while for the decrease of τ_0 the regular motion appears.

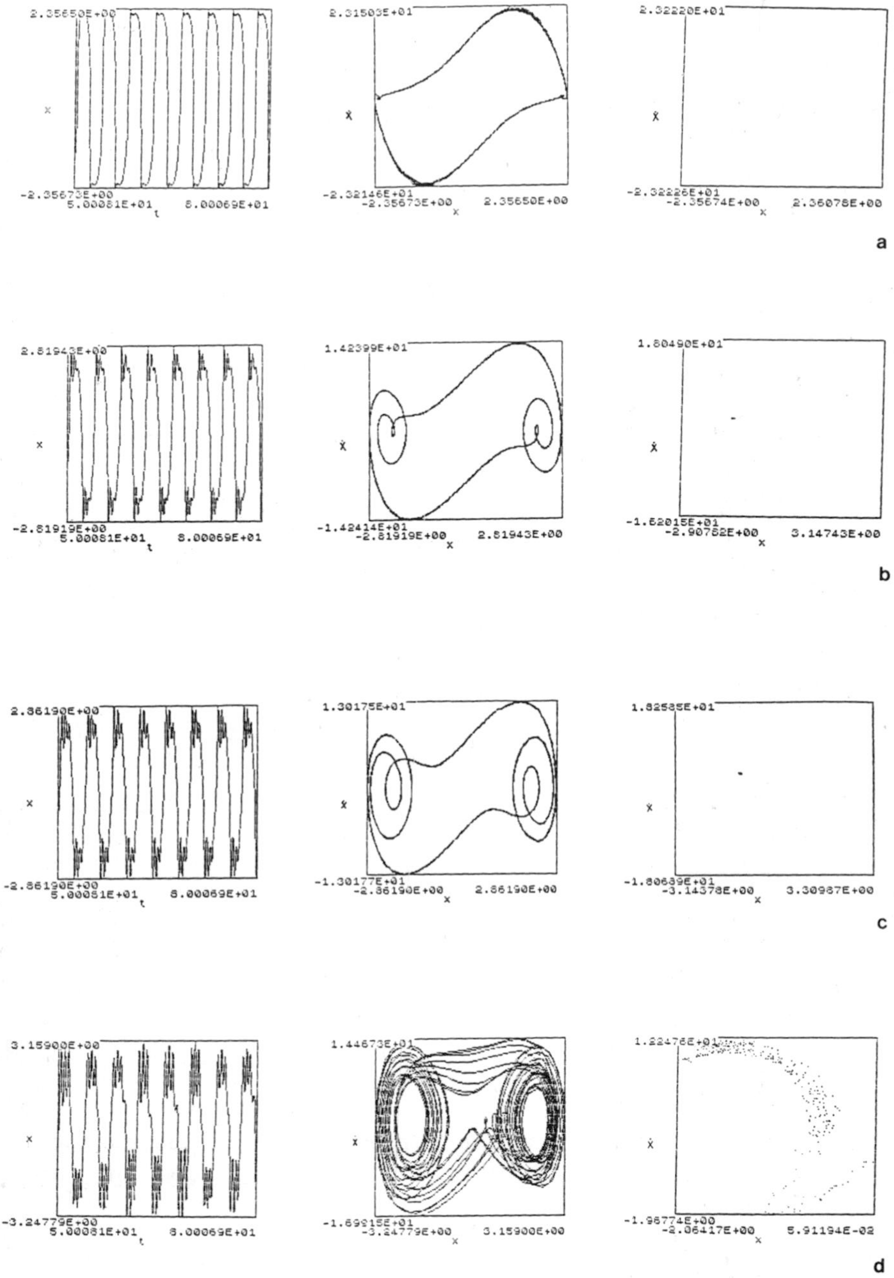
2.35650E+00
x
-2.35673E+00
5.00081E+01
t
8.00069E+01
2.31503E+01
ẋ
-2.32146E+01
-2.35673E+00
x
2.35650E+00
2.32220E+01
ẋ
-2.32226E+01
-2.35674E+00
x
2.36078E+00
a
2.81943E+00
x
-2.81919E+00
5.00081E+01
t
8.00069E+01
1.42399E+01
ẋ
-1.42414E+01
-2.81919E+00
x
2.81943E+00
1.80490E+01
ẋ
-1.62015E+01
-2.90752E+00
x
3.14743E+00
b
2.86190E+00
x
-2.86190E+00
5.00081E+01
t
8.00069E+01
1.30175E+01
ẋ
-1.30177E+01
-2.86190E+00
x
2.86190E+00
1.82585E+01
ẋ
-1.80689E+01
-3.14378E+00
x
3.30987E+00
c
3.15900E+00
x
-3.24779E+00
5.00081E+01
t
8.00069E+01
1.44673E+01
ẋ
-1.69215E+01
-3.24779E+00
x
3.15900E+00
1.22476E+01
ẋ
-1.96774E+00
-2.06417E+00
x
5.91194E-02
d

Fig. 23.

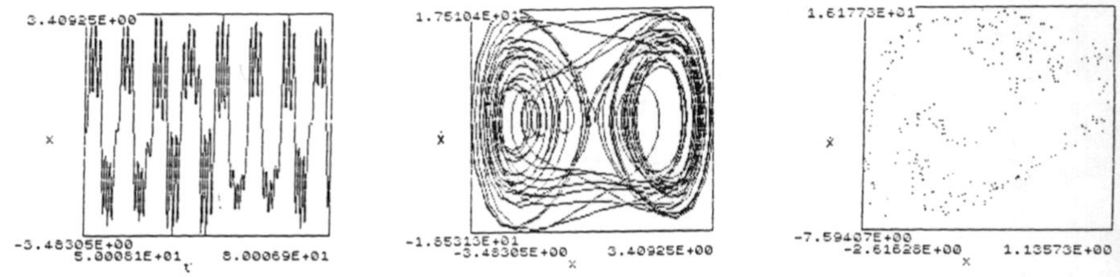

e

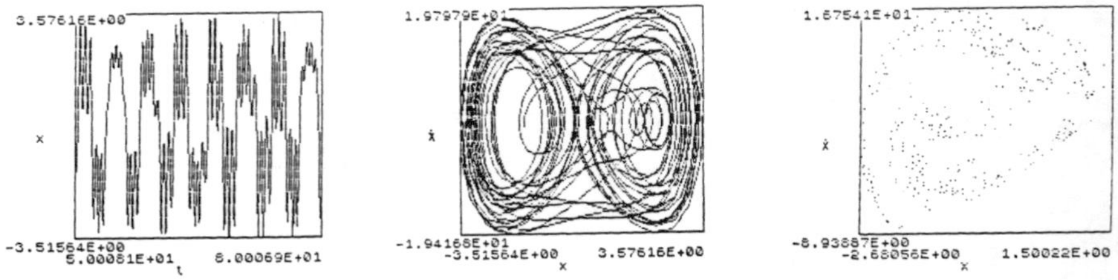

f

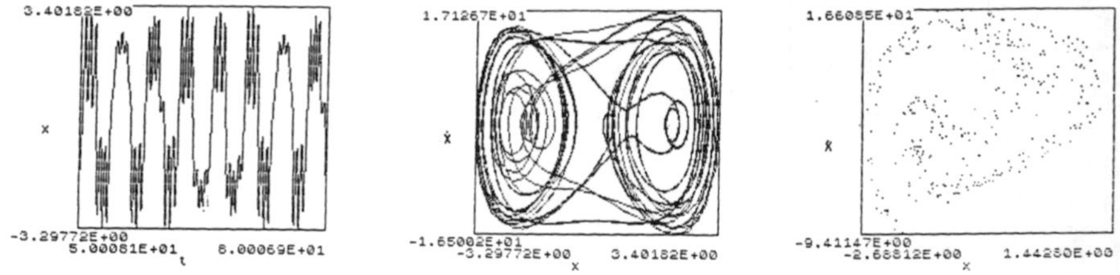

g

Fig. 23. Time histories, phase portraits and Poincaré maps for α: a) 8.0, b) 1.0, c) 0.5, d) 0.1, e) 0.01, f) 0.001, g) 0.0.

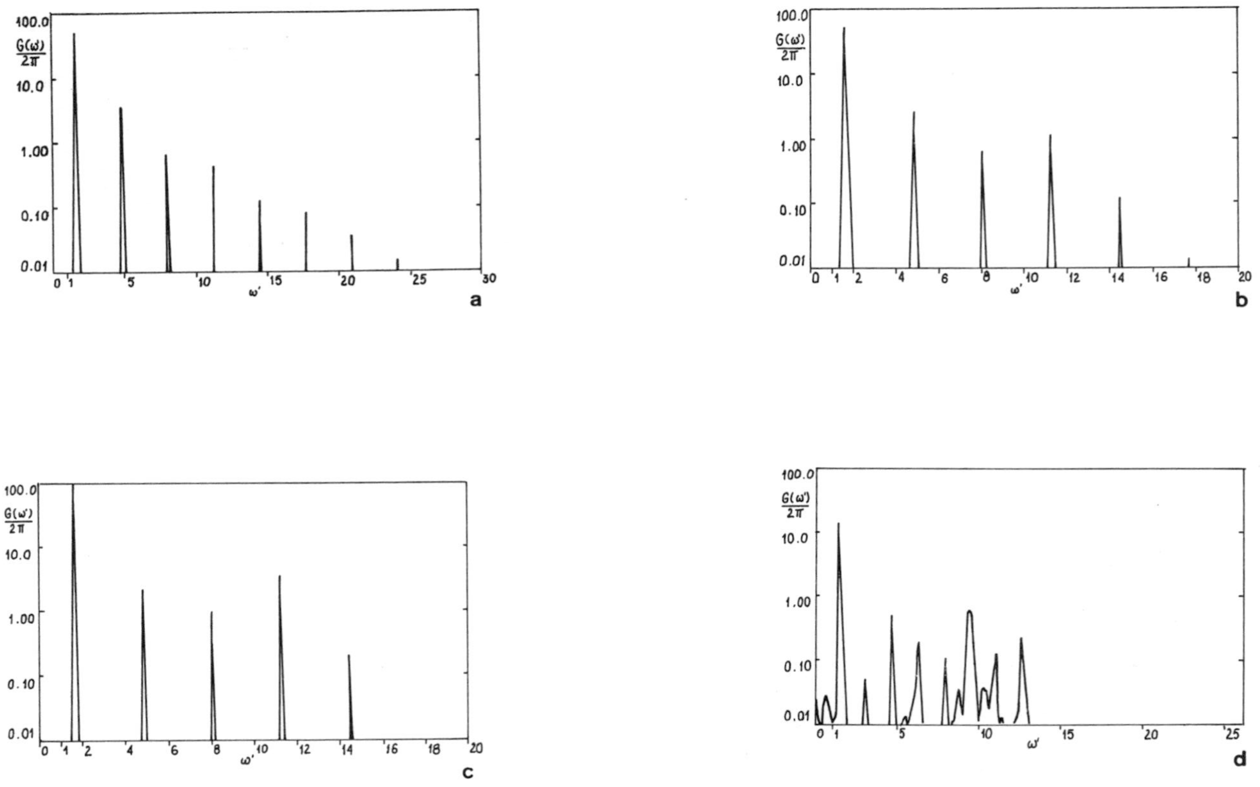

Fig. 24.

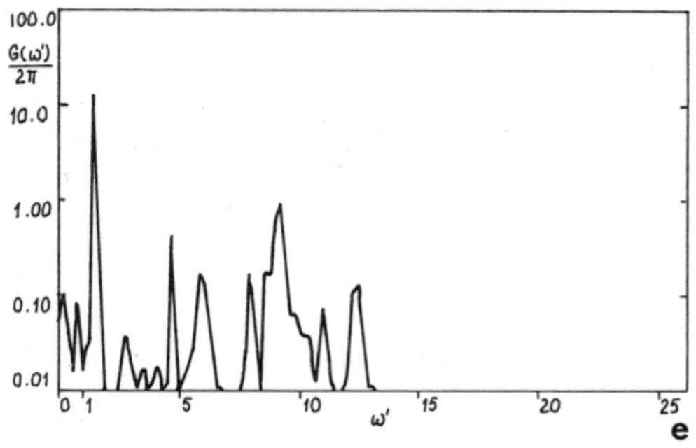

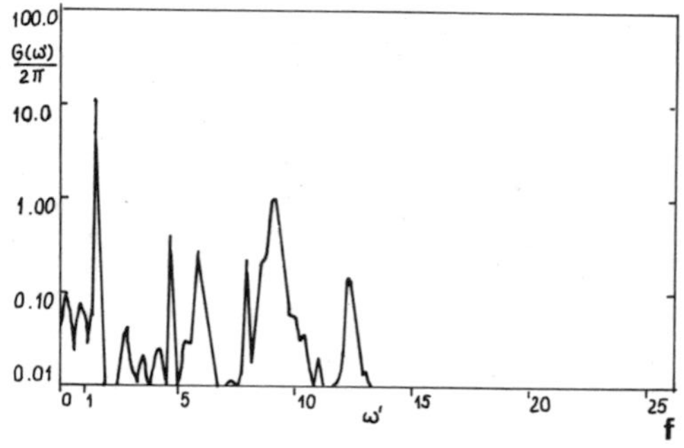

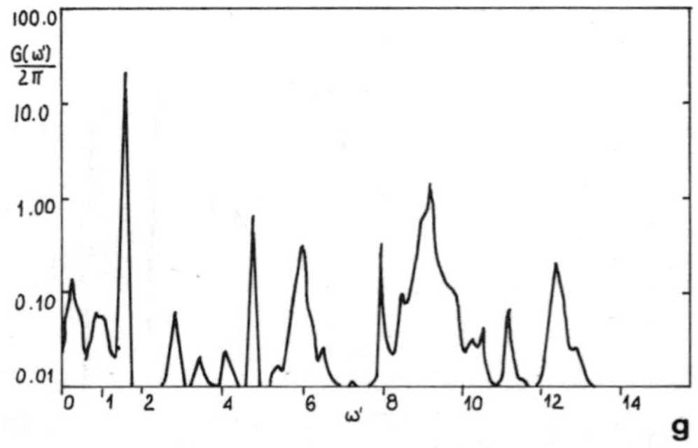

Fig. 24. Frequency spectra for the parameters as in Fig. 21.

3.4.2. Further observation of chaotic behaviour

For $\alpha = 0.2$, $\beta = 1.0$, $F = 17.0$, $\omega = 4.0$ and $k = 0.0$ Ueda and Akamatsu [29] have shown that oscillator (119) has chaotic orbits. Let us now investigate the behaviour of the strange attractor associated with the changes of the delay coefficients k and τ_0.

A. <u>A route to chaos via successive bifurcation of frequency</u>

We now observe the development of behaviour of the oscillator for $k = 10.0$ with the increase of the delay argument τ_0 (other parameters are fixed). For $\tau_0 = 1.0$ we have obtained regular motion. For this case the Poincaré map contains the points lying on the regular closed curve and the corresponding frequency spectra are discrete (Fig. 25a). For $\tau_0 = 1.5$ the periodic motion has appeared. This motion includes the harmonic frequency of the excited force and the third and fifth ultraharmonics. Then, for $\tau_0 = 1.75$, the motion becomes quasiperiodic again with the characteristic three groups of frequency. With the further increase of τ_0 (Fig. 25d) a "weak" chaos is exhibited. The points on the Poincaré map appear in an irregular way and the previously discrete values in the Fourier spectra tend to broaden slightly.

The strange attractor is then detected (Fig. 25e) with the

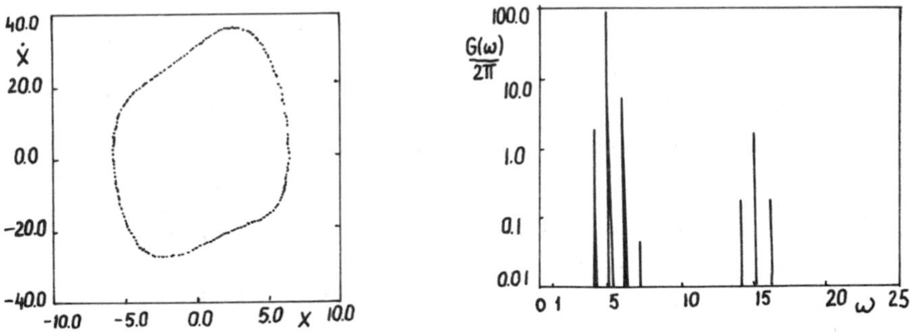

a

Fig. 25.

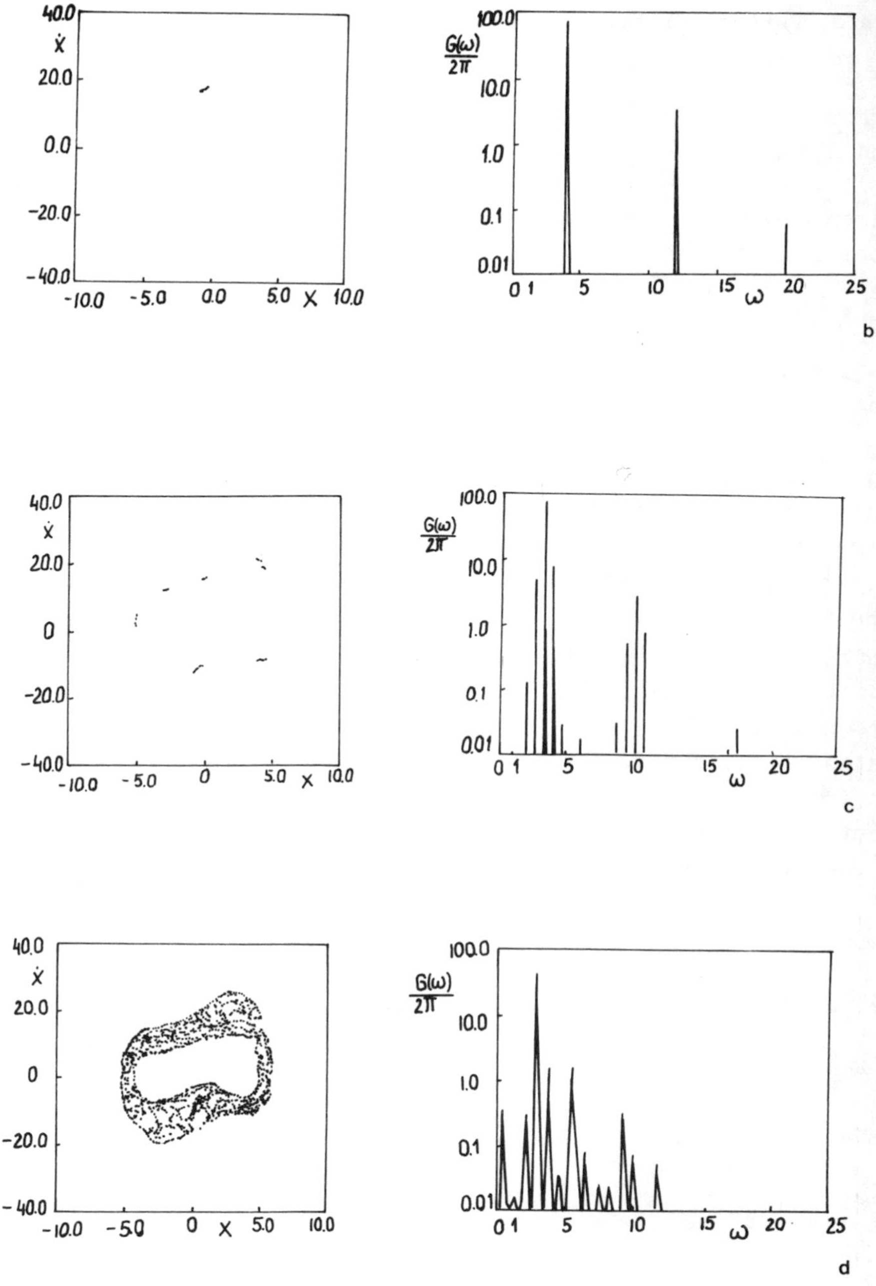

40.0
Ẋ
20.0
0.0
-20.0
-40.0
-10.0
-5.0
0.0
5.0
X
10.0
100.0
G(ω)/2π
10.0
1.0
0.1
0.01
0 1
5
10
15
ω
20
25
b
40.0
Ẋ
20.0
0
-20.0
-40.0
-10.0
-5.0
0
5.0
X
10.0
100.0
G(ω)/2π
10.0
1.0
0.1
0.01
0 1
5
10
15
ω
20
25
c
40.0
Ẋ
20.0
0
-20.0
-40.0
-10.0
-5.0
0
X
5.0
10.0
100.0
G(ω)/2π
10.0
1.0
0.1
0.01
0 1
5
10
15
ω
20
25
d

Fig. 25.

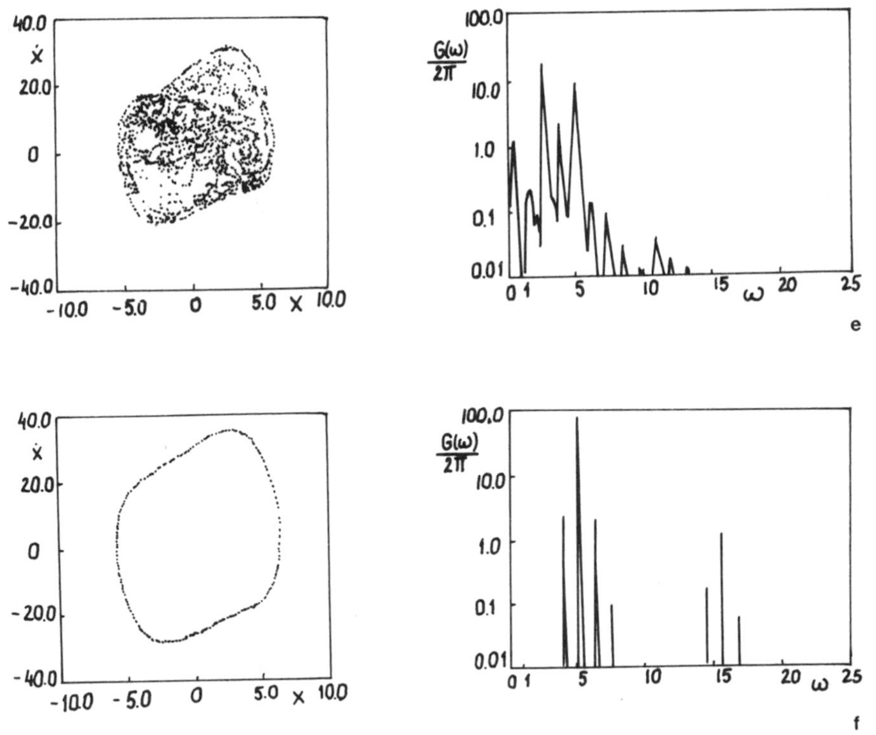

Fig. 25. Poincaré maps (left) and frequency spectra (right) for k = 10.0, (T_{min} = 250.0, T_{max} = 3000.0); a) τ_0 = 1.0, b) τ_0 = 1.5, c) τ_0 = 1.75, d) τ_0 = 1.9, e) τ_0 = 2.0, f) τ_0 = 2.15.

characteristic structure of the Poincaré map, the Fourier spectra having infinitely many components. The irregular motion exists however in a relatively small interval of τ_0. For τ_0 = 2.15 the motion is almost periodic again (Fig. 25f).

In Figure 26 a part of the "bifurcation tree" is presented, illustrating the transition from the regular motion to chaos, and then to regular again.

In this case we have shown that a further strange attractor exists near to Ueda's for k = 0.0. We have also demonstrated the transition from regular quasiperiodic motion via chaos to

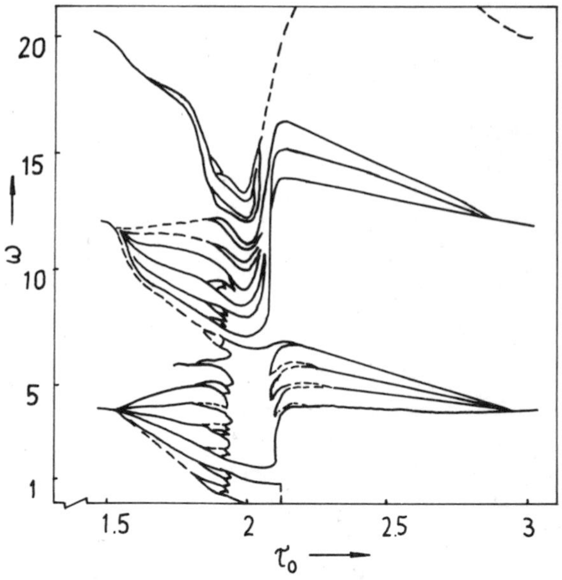

Fig. 26. Frequencies versus time delay τ_0 for the parameters as in Fig. 25.

an another regular quasiperiodic motion with increasing delay argument τ_0.

B. A route from chaos to regular motion

Now we consider the change of the strange attractor (for fixed value $\tau_0 = 1.0$) with increase of amplification coefficient k. From k = 0.0 to k = 0.1 the strange attractor becomes chaotic (see Figs. 27a,b). In both cases presented (for k = 0.001 and k = 0.01) the broad-band Fourier spectra lie on the left side of the fundamental frequency $\omega = 4$. For k = 0.1 the strange chaotic attractor suddenly disappears and periodic motion occurs with a fundamental frequency (i. e. corresponding to exciting force) and one subharmonic. For k = 2.0 we present the interesting result that, even in the strong nonlinear oscillator, the periodic orbit with one frequency (equal to ω) can appear. This harmonic motion remains unchang-

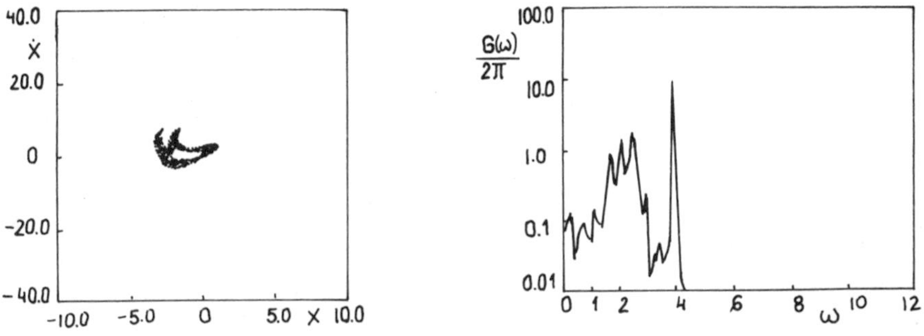
40.0
ẋ
20.0
0
-20.0
-40.0
-10.0 -5.0 0 5.0 x 10.0
100.0
G(ω)/2π
10.0
1.0
0.1
0.01
0 1 2 4 6 8 ω 10 12

a

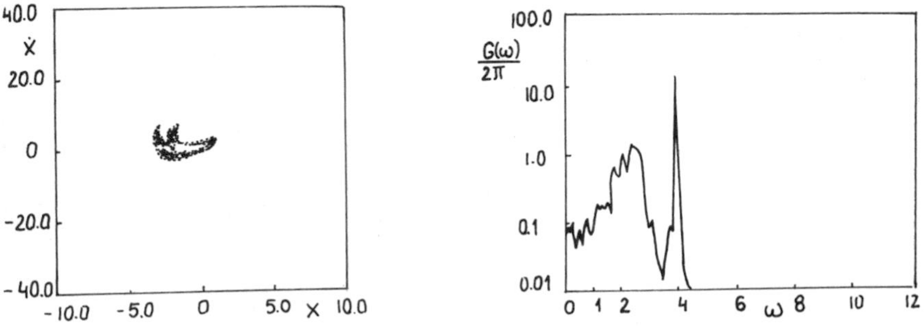
40.0
ẋ
20.0
0
-20.0
-40.0
-10.0 -5.0 0 5.0 x 10.0
100.0
G(ω)/2π
10.0
1.0
0.1
0.01
0 1 2 4 6 ω 8 10 12

b

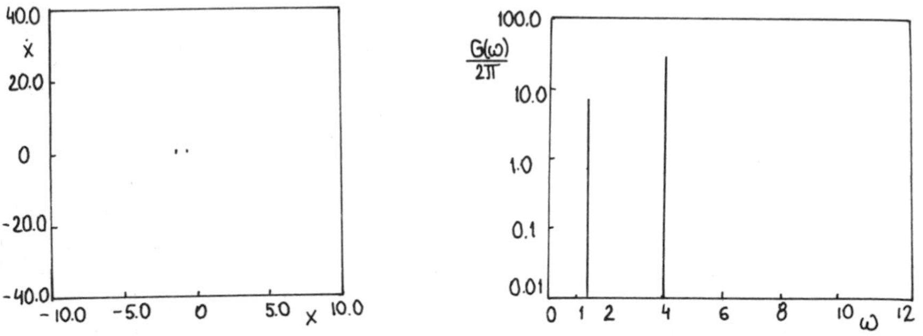
40.0
ẋ
20.0
0
-20.0
-40.0
-10.0 -5.0 0 5.0 x 10.0
100.0
G(ω)/2π
10.0
1.0
0.1
0.01
0 1 2 4 6 8 10 ω 12

c

Fig. 27.

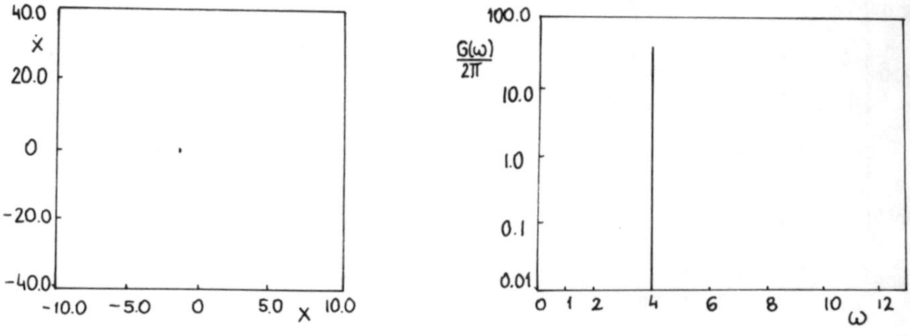
40.0
Ẋ
20.0
0
-20.0
-40.0
-10.0 -5.0 0 5.0 X 10.0
100.0
G(ω)/2π
10.0
1.0
0.1
0.01
0 1 2 4 6 8 10 12
ω

d

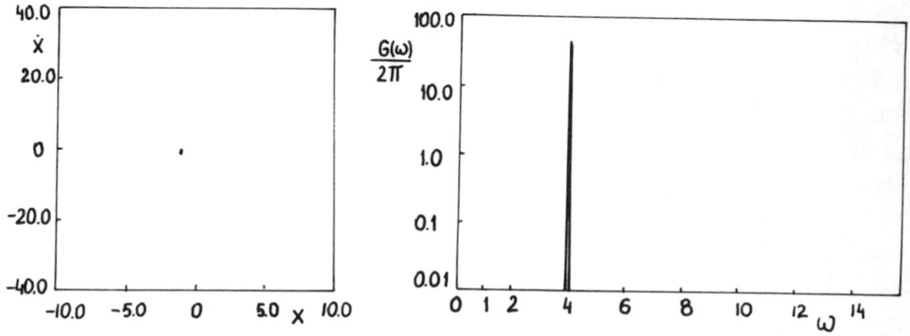
40.0
Ẋ
20.0
0
-20.0
-40.0
-10.0 -5.0 0 5.0 X 10.0
100.0
G(ω)/2π
10.0
1.0
0.1
0.01
0 1 2 4 6 8 10 12 14
ω

e

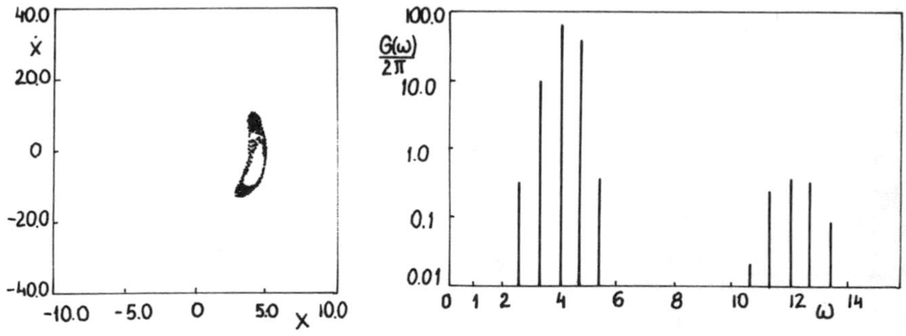
40.0
Ẋ
20.0
0
-20.0
-40.0
-10.0 -5.0 0 5.0 X 10.0
100.0
G(ω)/2π
10.0
1.0
0.1
0.01
0 1 2 4 6 8 10 12 14
ω

f

Fig. 27.

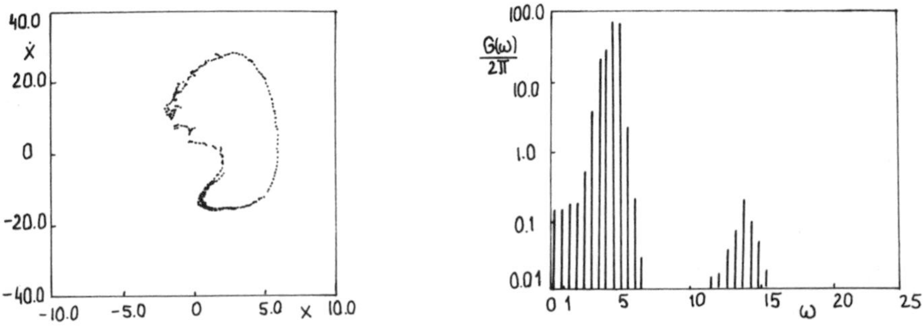

g

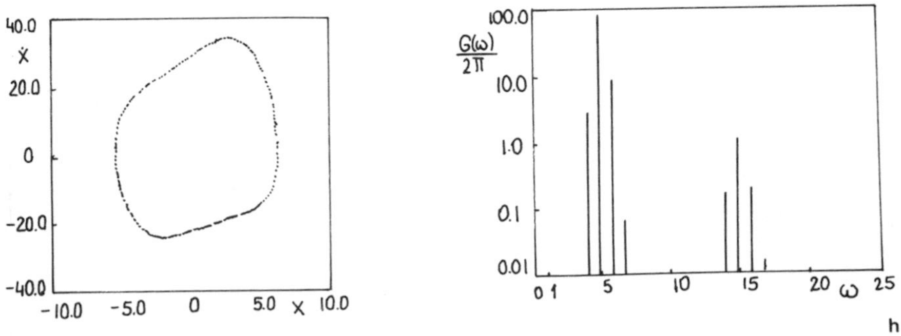

h

Fig. 27. Poincaré maps (left) and frequency spectra (right) for $\tau_0 = 1.0$ ($T_{min} = 39.3$, $T_{max} = 2000.0$); a) k = 0.001, b) k = 0.01, c) k = 0.1, d) k = 2.0, e) k = 3.0, f) k = 6.75, g) k = 7.25, h) k = 9.0.

ed with the further increase of τ_0. In the interval $6.75 \leqslant k \leqslant 7.25$, attractors are observed whose orbits show very complicated motion.

C. Periodic window with two frequencies

Let us observe the influence of changing the delay argument τ_0 for the fixed value k = 0.1 (Fig. 28). For $\tau_0 = 2.0$ the strange chaotic attractor appears (Fig. 28a). Chaotic orbital states remain up to $\tau_0 = 5.0$. For $\tau_0 = 5.0$ periodic motion

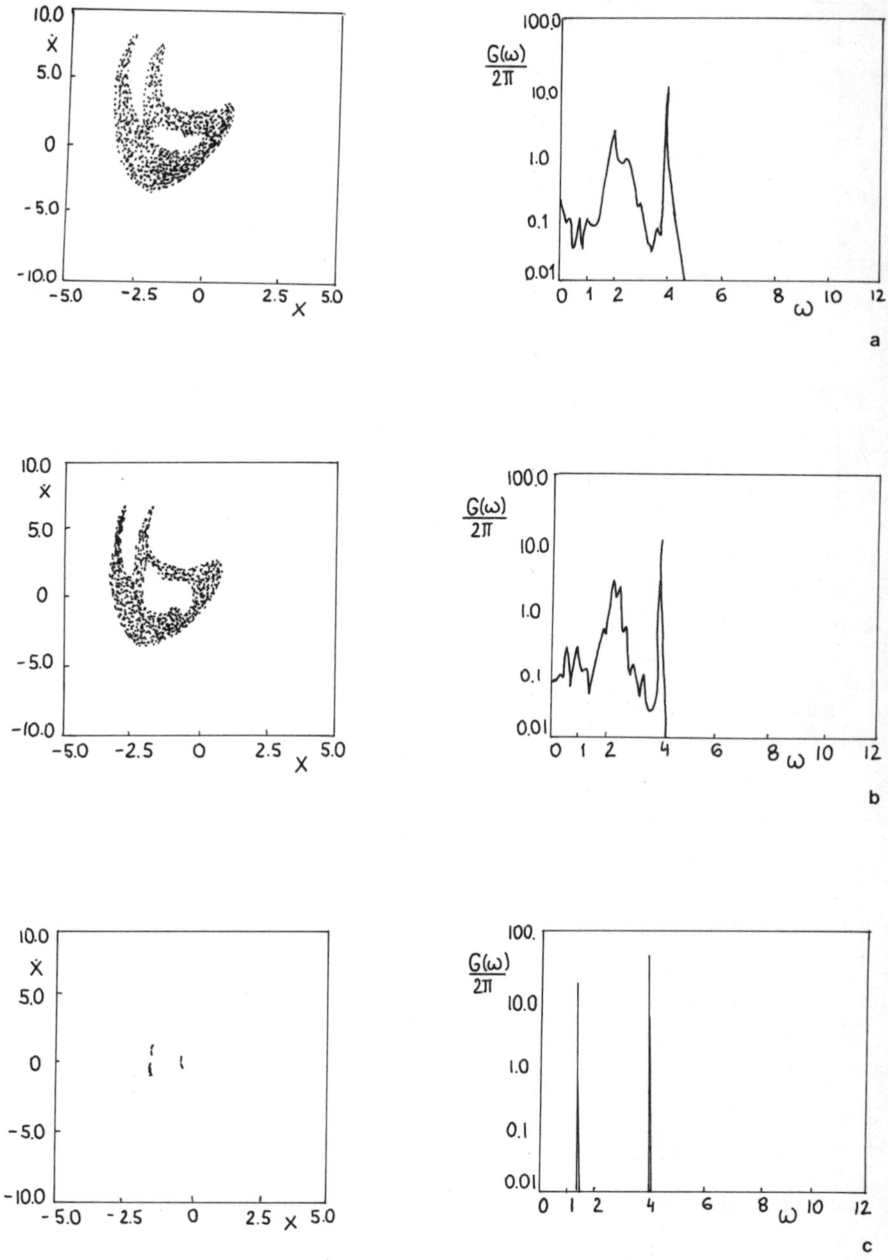
10.0
Ẋ
5.0
0
-5.0
-10.0
-5.0
-2.5
0
2.5
X
5.0
100.0
G(ω)/2π
10.0
1.0
0.1
0.01
0
1
2
4
6
8
ω
10
12
a
b
c

Fig. 28.

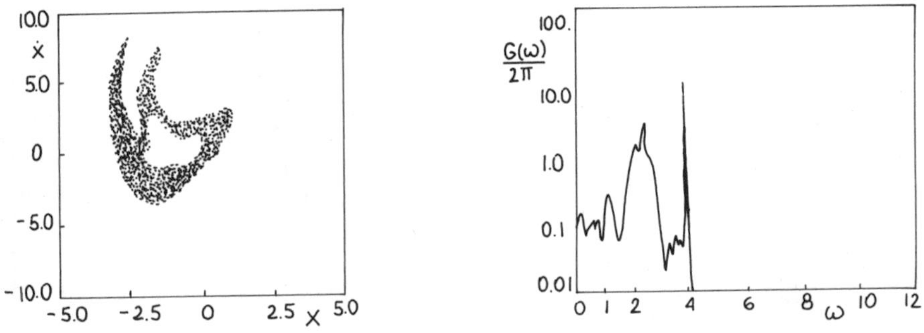

d

Fig. 28. Poincaré maps (left) and frequency spectra (right) for k = 0.1 (T_{min} = 39.3, T_{max} = 3000.0); a) τ_0 = 2.0, b) τ_0 = 3.0, c) τ_0 = 5.0, d) τ_0 = 10.0.

with two frequencies has appeared. Then chaotic motion is observed again (Fig. 28d).

In the τ_0 interval considered here the dominant motion was chaotic. We have found however, between the two types of strange attractors, periodic motion with fundamental and one subharmonic frequency.

D. Periodic window with one frequency

Now, our procedure will be analogous to the case described

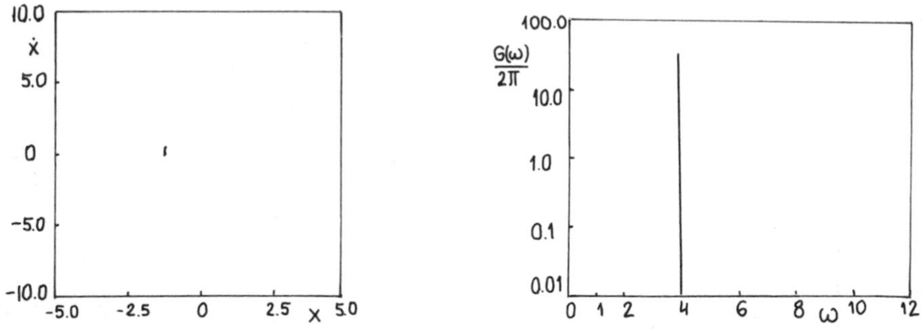

a

Fig. 29.

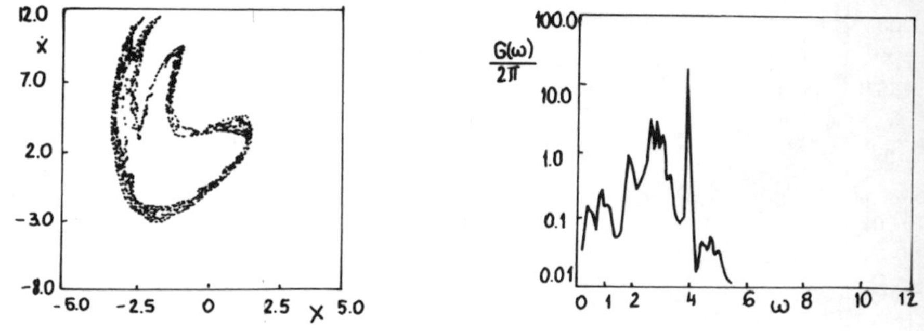
12.0
ẋ
7.0
2.0
-3.0
-8.0
-5.0 -2.5 0 2.5 X 5.0
100.0
G(ω)/2π
10.0
1.0
0.1
0.01
0 1 2 4 6 ω 8 10 12

b

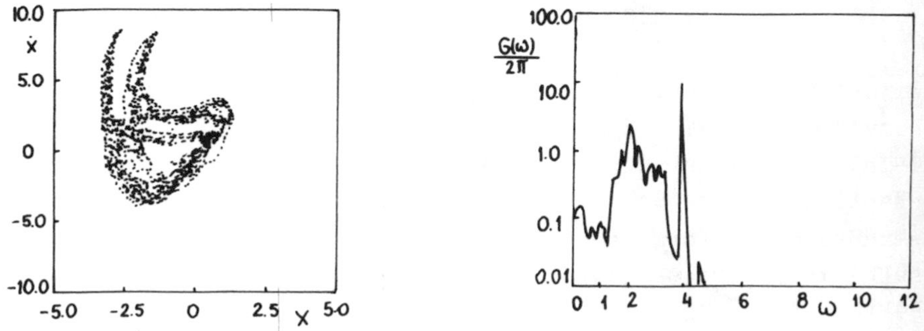
10.0
ẋ
5.0
0
-5.0
-10.0
-5.0 -2.5 0 2.5 X 5.0
100.0
G(ω)/2π
10.0
1.0
0.1
0.01
0 1 2 4 6 8 ω 10 12

c

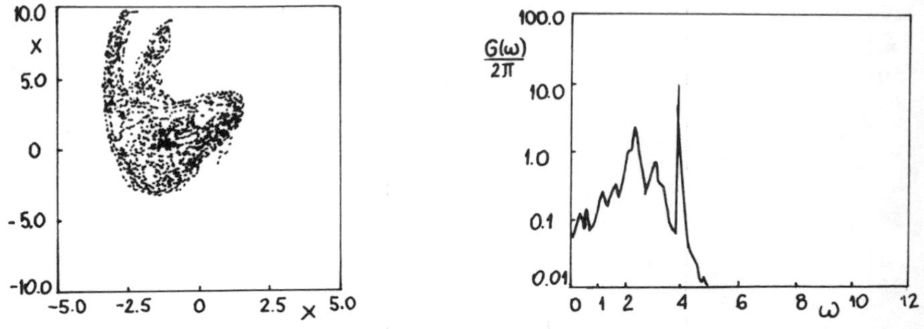
10.0
X
5.0
0
-5.0
-10.0
-5.0 -2.5 0 2.5 X 5.0
100.0
G(ω)/2π
10.0
1.0
0.1
0.01
0 1 2 4 6 8 ω 10 12

d

Fig. 29.

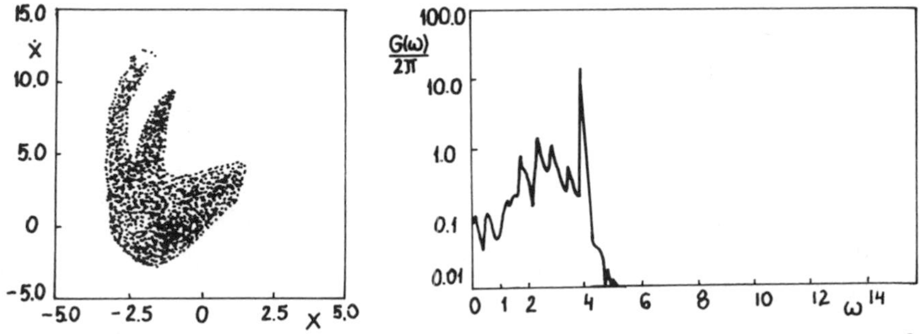

e

Fig. 29. Poincaré maps (left) and frequency spectra (right) for k = 1.0 (T_{min} = 39.3, T_{max} = 3000.0); a) τ_0 = 1.0, b) τ_0 = 2.0, c) τ_0 = 3.0, d) τ_0 = 5.0, e) τ_0 = 10.0.

in section C, but the fixed value is k = 1.0. For τ_0 = 1.0, even though the oscillator is strongly nonlinear, the motion is harmonic with the frequency of exciting force (Fig. 29a). For τ_0 = 2.0 (Fig. 29b) chaotic orbits occur. The continuous part of the Fourier spectra lies left of $\omega = 4$. With increasing τ_0, chaotic behaviour of the oscillator is more evident (Fig. 29c, d, e).

References

1. Butenin, I. W., Neimark, I. and Fufaiev, N. A. "Introduction in the Theory of Nonlinear Oscillations". Nauka: Moscow (1976), in Russian.
2. Marsden, J. E. and McCracken, M., Appl. Math. Sci. 19, "The Hopf Bifurcations and Its Applications". New York: Springer (1976).
3. Iooss, G. and Joseph, D. D., "Elementary Stability and Bifurcation Theory". Berlin, Heidelberg, New York: Springer (1980).
4. Hassard, B. D., Kazarinoff N. D. and Wan Y. H., "Theory and Applications of Hopf Bifurcation". Cambridge: University Press (1981).
5. Malkin, I. G., "Some Problems in the Theory of Nonlinear Oscillations". Moscow (1956), in Russian.
6. Iakubovich, V. A. and Starzhinskii, V. M., "Linear Differential Equations with Periodic Coefficients and their Applications". Nauka: Moscow (1972), in Russian.
7. Giacaglia, G. E., "Perturbations Methods in Nonlinear Systems". New York: Springer (1972).
8. Nayfeh, A. H. and Mook, P. T., "Nonlinear Oscillations". New York: John Wiley (1979).
9. Nayfeh, A. H., "Introduction to Perturbation Techniques". New York: John Wiley (1981).
10. Carr, J., "Applications of Center Manifold Theory". New York: Springer (1981).
11. Huseyin, K. and Atadan, A. S., "On the Analysis of the Hopf Bifurcation", Int. J. Engng Sci. 21, 247-262 (1983).
12. Atadan, A. S. and Huseyin, K., "On the Oscillatory Instability of Multiple - Parameter Systems", Int. J. Engng Sci. 23(8), 857-873 (1985).
13. Yu, P. and Huseyin, K., "Static and Dynamic Bifurcations Associated with a Double Zero Eigenvalue", Dyn. Stab. Sys. 1, 73-86 (1986).

14. Yu, P. and Huseyin, K., "Bifurcations Associated with a Double Zero and a Pair of Pure Imaginary Eigenvalues", SIAM J. Appl. Math. 48(2), 229-261 (1988).
15. Rosenblat, S. and Cohen, D. S., "Periodically Perturbated Bifurcation - II. Hopf Bifurcation", Stud. Appl. Math. 64, 143-175 (1981).
16. Kath, W. L., "Resonance in Periodically Perturbated Hopf Bifurcation", Stud. Appl. Math. 65, 95-112 (1981).
17. Bajaj, A. K., "Resonant Parametric Perturbations of the Hopf Bifurcation", J. Math. Anal. Appl. 115, 214-224 (1986).
18. Sri Namachchivaya, N. and Ariaratnam, S. T., "Periodically Perturbated Hopf Bifurcation", SIAM J. Appl. Math. 47(1), 15-39 (1987).
19. Awrejcewicz, J., "On the Hopf Bifurcation", Nonlinear Vibr. Problems, Polish Academy of Science (to appear).
20. Awrejcewicz, J., "Analysis of the Biparameter Hopf Bifurcation", Nonlinear Vibr. Problems, Polish Academy of Science (to appear).
21. Awrejcewicz, J., "Hopf Bifurcation in Mathieu-Duffing's Oscillator", Nonlinear Vibr. Problems, Polish Academy of Science (to appear).
22. Awrejcewicz, J., "Analysis of the Double Hopf Bifurcation", Nonlinear Vibr. Problems, Polish Academy of Science (to appear).
23. Awrejcewicz, J., "Hopf Bifurcation in Duffing's Oscillator", Proc. PAN American Congress of Appl. Mech., Rio de Janeiro, Brazil, Jan. 3-6, 640-643 (1989).
24. Awrejcewicz, J., "An Analytical Method for Detecting Hopf Bifurcation Solutions in Non-Stationary Non-Linear Systems", J. Sound Vibr. 129(1), 175-178 (1989).
25. Awrejcewicz, J., "System Vibrations: Rotor with Self-Excited Support", Proc. Intern. Conf. Rotordyn. in Tokyo, Sept. 14-17, 517-522 (1986).
26. Awrejcewicz, J., "Determination of the Limits of the Unstable Zones of the Unstationary Non-Linear Mechanical Systems", Int. J. Nonlinear Mech. 23(1), 87-94(1988).
27. Lorenz, E. N., "Deterministic Nonperiodic Flow", J. Atm.

Sci. 20, 130-141 (1963).

28. Ueda, Y., "Randomly Transitional Phenomena in the System Governed by Duffing's Equation", J. Stat. Phys. 20, 181-196 (1979).
29. Ueda, Y. and Akamatsu, N., "Chaotically Transitional Phenomena in the Forced Negative - Resistance Oscillator", IEEE Trans. Circ. Sys. 28, 217-223 (1981).
30. Raty, R., Isomäki, H. M. and Boehm, J., "Chaotic Motion of a Classical Anharmonic Oscillator", Acta Pol. Scand. Me 85, 1-30 (1984).
31. Holmes, P., "A Nonlinear Oscillator with a Strange Attractor", Phil. Trans. R. Soc., London Ser. A 292, 419-448 (1979).
32. Troger, H., "Chaotic Behaviour in Simple Mechanical Systems", ZAMM 62, 18-27 (1982).
33. Popp, R., "Chaotic Motion in Duffing's Oscillator". Festschrift für K. Magnus, München, 269-296 (1982) - in German.
34. Awrejcewicz, J., "Chaos in Simple Mechanical Systems with Friction", J. Sound Vibr. 109(1), 178-180 (1986).
35. Kapitaniak, T., Awrejcewicz, J. and W. - H. Steeb, "Chaotic Behaviour of an Anharmonic Oscillator with Almost Periodic Excitation", J. Phys. A: Math. Gen. 20, L355-L358 (1987).
36. Rössler, O. E., private communication.
37. Brommundt, E., "On the Numerical Investigation of Non-linear Periodic Rotor Vibrations"; in Dynamics of Rotors, IUTAM - Symposium in Lyngby (Denmark). Berlin, Heidelberg, New York: Springer, 75-102 (1975).
38. Brommundt, E., "Bifurcation of Self-Excited Rotor Vibrations", VII Intern. Conf. Nonl. Vibr. Berlin: Akademie - Verlag, 123-134 (1977).
39. Urabe, M., "Galerkin's Procedure for Nonlinear Periodic Systems", Arch. Rat. Mech. Anal. 20, 120-152 (1965).
40. Urabe, M. and Reiter, A., "Numerical Computation of Non-linear Forced Oscillations by Galerkin's Procedure", J. Math. Analysis Appl. 14, 107- 140 (1966).
41. Seydel, R., "Numerical Computation of Periodic Orbits

that Bifurcate from Stationary Solutions of Ordinary Differential Equations", Appl. Math. Comput. 9, 257-271 (1981).

42. Seydel, R. and Hlavacek, V., "Strategy of Calculation of Periodic Solutions", Chem. Eng. Sci. 42(12), 2927-2933 (1987).
43. Seydel, R., "New Methods for Calculating Stability of Periodic Solutions", Comput. Math. Applic. 14, 505-510 (1987).
44. Seydel, R., "From Equilibrium to Chaos: Practical Bifurcation and Stability Analysis". New York: Elsevier (1988).
45. Ueda, Y., "Explosion of Strange Attractors Exhibited by Duffing's Equation", Ann. New York Acad. Sci. 357, 422-423 (1980).
46. Szemplinska-Stupnicka, W. and Bajkowski, J., "The 1/2 Subharmonic Resonance and Its Transition to Chaotic Motion in Nonlinear Oscillator", Int. J. Non-Lin. Mech. 21, 401-419 (1986).
47. Guckenheimer, J. and Holmes, P., "Nonlinear Oscillations, Dynamical Systems and Bifurcations of Vector Fields". New York: Springer (1983).
48. Szemplinska-Stupnicka, W., "The Refined Approximate Criterion for Chaos in a Two State Mechanical Oscillator", Ing. Arch. 58, 354-366 (1988).
49. Feigenbaum, M. J., "The Universal Metric Properties of Nonlinear Transformations", J. Stat. Phys. 21(6), 669-706 (1979).
50. Tousi, S. and Bajaj, A. K., "Period - Doubling Bifurcations and Modulated Motions in Forced Mechanical Systems" Trans. ASME J. Appl. Mech. 52, 446-452 (1985).
51. Korn, G. A. and Korn, T. M., "Mathematical Handbook for Scientists and Engineers". New York, Toronto, London: McGraw Book Company (1961).
52. Arnold, V. I., "Geometrical Methods in the Theory of Ordinary Differential Equations". New York, Heidelberg, Berlin: Springer (1983).
53. Aronson, P. G., Chory, M. A., Hall, G. R. and McGehee,

R. P., "Bifurcations from an Invariant Circle for Two Parameter Families of Maps of the Plane: A Computer Assisted Study", Comm. Math. Phys. 83, 303-354 (1981).

54. Cronjaeger, R., "Model of the Sound Generation in a Human Larynx", PhD Thesis, Braunschweig 1978 (not published) - in German.

55. Awrejcewicz, J., "Numerical Investigations of the Constant and Periodic Motions of the Human Vocal Cords Including Stability and Bifurcation Phenomena", Int. J. Dyn. Stab. Sys. (to appear).

56. Guckenheimer, J., "Persistent Properties of Bifurcations", Physica 7D, 105-110 (1983).

57. Awrejcewicz, J. and Grabski, J., "Chaos in a Particular Nonlinear Oscillator", Acta Mechanica (to appear).

58. Awrejcewicz, J., "Two Kinds of Evolution of Strange Attractors for the Example of a Particular Nonlinear Oscillator", J. Appl. Math. Phys. ZAMP 40, 375-386(1989).

59. Awrejcewicz, J. and Mrozowski, J., "Bifurcations and Chaos of a Particular Van der Pol - Duffing's Oscillator" J. Sound Vibr. (to appear).

60. Awrejcewicz, J., "A Route to Chaos in a Nonlinear Oscillator with Delay", Acta Mechanica 77, 111-120 (1989).

61. Awrejcewicz, J. and Wojewoda, J., "Observation of Chaos in a Nonlinear Oscillator with Delay: A Numerical Study", KSME Journal (to appear).

62. Awrejcewicz, J., "Chaotic Motion in a Nonlinear Oscillator with Friction", KSME Journal 2(2), 104-109 (1988).

63. Arrowsmith, P. and Taha, K., "Bifurcations of a Particular Van der Pol Oscillator", Meccanica 18, 195-204 (1983).

64. Grebogi, C. and Ott, E., "Crises, Sudden Changes in Chaotic Attractors and Transient Chaos", Physica 7D, 181-200 (1983).

65. Hall, G. and Watt, J. M., "Modern Numerical Methods of Ordinary Differential Equations". Oxford: Clarendon Press (1976).

66. Curry, J. and Yorke, J., "A Transition from Hopf Bifurcation to Chaos: A Computer Experiments on Maps in R^2. The Structure of Attractors in Dynamical Systems".

Subject Index